Directed Reading A

Section: Characteristics of the Atmosphere

Write the letter of the correct answer in the space provided.

_______ **1.** What is Earth's atmosphere?
- **a.** oxygen and sunlight
- **b.** a mixture of particles
- **c.** a mixture of gases
- **d.** water vapor

THE COMPOSITION OF THE ATMOSPHERE

_______ **2.** Earth's atmosphere is made up mostly of which gas?
- **a.** oxygen
- **b.** nitrogen
- **c.** carbon dioxide
- **d.** argon

_______ **3.** How much of the atmosphere is oxygen?
- **a.** about 1%
- **b.** about 78%
- **c.** about 20%
- **d.** about 100%

_______ **4.** Where is most of the water in the atmosphere?
- **a.** in rain
- **b.** in ice
- **c.** in water vapor
- **d.** in carbon dioxide

AIR PRESSURE AND TEMPERATURE

_______ **5.** At sea level, a square inch of Earth is under how much air?
- **a.** 150 lbs
- **b.** 15 lbs
- **c.** 30 lbs
- **d.** 1500 lbs

| Directed Reading A *continued*

Altitude and Air Pressure

_______ **6.** What pulls the gas molecules in the atmosphere toward Earth's surface?
 a. air pressure
 b. gravity
 c. water
 d. solar energy

_______ **7.** What is the measure of the force with which air molecules push on a surface?
 a. Earth's surface
 b. altitude
 c. water vapor
 d. air pressure

_______ **8.** Where is air pressure strongest?
 a. on a mountain top
 b. at Earth's surface
 c. in outer space
 d. close to the sun

Atmospheric Composition and Air Temperature

_______ **9.** Which of the following do warmer layers of the atmosphere have more of than cold layers?
 a. gases that make heat
 b. gases that absorb water
 c. gases that absorb cold
 d. gases that absorb energy from the sun

LAYERS OF THE ATMOSPHERE
The Troposphere: The Layer in Which We Live

_______ **10.** Where does the troposphere lie?
 a. on mountaintops
 b. in the thermosphere
 c. next to Earth's surface
 d. at the ocean bottom

Directed Reading A *continued*

The Stratosphere: Home of the Ozone Layer

______ **11.** Where is the stratosphere?
    a. within the troposphere
    b. below the troposphere
    c. above the troposphere
    d. below the mesosphere

______ **12.** What does the ozone layer absorb to protect life?
    a. gas mixtures
    b. ultraviolet rays
    c. particle mixtures
    d. water vapor

The Mesosphere: The Middle Layer

______ **13.** As altitude increases in the mesosphere, what does temperature do?
    a. It increases.
    b. It decreases.
    c. It remains unchanged.
    d. It cannot be measured.

The Thermosphere: The Edge of the Atmosphere

______ **14.** What do nitrogen and oxygen atoms in the thermosphere do to solar radiation?
    a. absorb it
    b. emit it
    c. reflect it
    d. destroy it

Directed Reading A

Section: Atmospheric Heating

RADIATION: ENERGY TRANSFER BY WAVES

Write the letter of the correct answer in the space provided.

_______ **1.** How long does it take the sun's energy to reach Earth?
   a. about 8 hours
   b. about 80 hours
   c. about 8 minutes
   d. about 8 days

_______ **2.** What is the transfer of energy as electromagnetic waves, including those hitting Earth, called?
   a. conduction
   b. radiation
   c. reflection
   d. absorption

The Electromagnetic Spectrum

_______ **3.** What is the name for all frequencies or wavelengths of electromagnetic waves?
   a. solar energy
   b. ultraviolet radiation
   c. electromagnetic spectrum
   d. reflected sunlight

_______ **4.** What makes the kinds of electromagnetic radiation differ?
   a. wavelength
   b. radiation
   c. absorption
   d. reflection

The Atmosphere and Solar Radiation

_______ **5.** What results when solar radiation reaches Earth's atmosphere?
   a. Some becomes absorbed.
   b. All becomes absorbed.
   c. None becomes absorbed.
   d. All becomes ozone.

_______ **6.** What is the form of most solar energy that reaches Earth's surface?
   a. ultraviolet waves
   b. visible light
   c. radio waves
   d. infrared radiation

Directed Reading A *continued*

CONDUCTION: ENERGY TRANSFER BY CONTACT

_______ **7.** How is heat transferred in conduction?
- **a.** by heat waves
- **b.** by heat energy
- **c.** by wavelengths
- **d.** by physical contact

_______ **8.** In conduction, where is the kinetic energy of atoms in a hot object transferred?
- **a.** to hot objects
- **b.** to cold objects
- **c.** to potential energy
- **d.** to solar energy

CONVECTION: ENERGY TRANSFER BY MOTION

_______ **9.** How does convection cause heat transfer to take place in a liquid or gas?
- **a.** by radiation
- **b.** by evaporation
- **c.** by conduction
- **d.** by circulation

_______ **10.** What is the cycle of warm fluid rising and cool fluid sinking called?
- **a.** radiation current
- **b.** convection current
- **c.** conduction current
- **d.** density current

THE GREENHOUSE EFFECT

_______ **11.** What is the process where atmospheric gases absorb and reradiate solar energy called?
- **a.** thermal recycling
- **b.** thermal balance
- **c.** global warming
- **d.** greenhouse effect

The Radiation Balance: Energy In, Energy Out

_______ **12.** How does the radiation balance affect Earth?
- **a.** It causes global warming.
- **b.** It prevents air pollution.
- **c.** It makes Earth livable.
- **d.** It circulates solar energy.

GLOBAL WARMING

_______ **13.** What is one possible cause of increased average global temperatures?
- **a.** radiating solar energy
- **b.** circulating thermal energy
- **c.** burning fossil fuels
- **d.** reducing greenhouse gases

Directed Reading A

Section: Air Movement and Wind

WHAT CAUSES WIND?

Write the letter of the correct answer in the space provided.

______ **1.** What causes differences in air pressure?
- **a.** unequal heating of Earth
- **b.** unequal oxygen on Earth
- **c.** equal heating of Earth
- **d.** equal temperature of Earth

______ **2.** What causes wind?
- **a.** differences in water
- **b.** differences in gases
- **c.** differences in air pressure
- **d.** differences in oxygen

______ **3.** What are the large, circular patterns air travels in called?
- **a.** pressure belts
- **b.** pressure cells
- **c.** convection belts
- **d.** convection cells

The Coriolis Effect

______ **4.** What is the Coriolis effect, the curving of winds and ocean currents, due to?
- **a.** Earth's rotation
- **b.** Earth's surface area
- **c.** Earth's gravity
- **d.** Earth's atmosphere

______ **5.** In what hemisphere does the Coriolis effect cause northbound winds to travel east?
- **a.** Southern
- **b.** Northern
- **c.** Eastern
- **d.** Western

Directed Reading A *continued*

GLOBAL WINDS

______ **6.** With what do convection cells, pressure belts, and winds combine to
form global winds?
a. Earth's gravity
b. Coriolis effect
c. thermal patterns
d. pressure patterns

LOCAL WINDS

Use the terms from the following list to complete the sentences below.

sea breeze	valley breeze	local wind
land breeze	mountain breeze	

7. Air flow caused by the properties of Earth's surface matter is

a(n) _____________________.

8. Cool air that flows over the ocean toward land during the day is called

a(n) _____________________.

9. Cool air that flows over the land toward the ocean during the night is

called a(n) _____________________.

10. Warm air that rises up mountain slopes during the day is called

a(n) _____________________.

11. Cool air that moves down mountain slopes during the night is

called a(n) _____________________.

Directed Reading A

Section: The Air We Breathe

AIR POLLUTION

Write the letter of the correct answer in the space provided.

______ **1.** What is the introduction of pollutants from human and natural sources into the atmosphere called?
   a. water pollution
   b. air pollution
   c. ground pollution
   d. thermal pollution

Primary Pollutants

______ **2.** What are pollutants that are put directly into the air by human or natural activity?
   a. secondary pollutants
   b. primary pollutants
   c. natural pollutants
   d. animal pollutants

Secondary Pollutants

______ **3.** What results when primary pollutants react with other primary pollutants or with naturally occurring substances?
   a. inferior pollutants
   b. reactant pollutants
   c. secondary pollutants
   d. original pollutants

______ **4.** What is an example of a secondary pollutant?
   a. oxygen
   b. vapor
   c. pollen
   d. ozone

The Formation of Smog

______ **5.** What do ozone and vehicle exhaust react with to form smog?
   a. sunlight
   b. ultraviolet energy
   c. dust particles
   d. smoke

Directed Reading A *continued*

HUMAN-CAUSED AIR POLLUTION

______ **6.** How much human-caused air pollution in the United States comes
from cars?
a. about 10% to 20%
b. about 25% to 35%
c. about 40% to 50%
d. about 60% to 70%

Industrial Air Pollution

______ **7.** What is a very big cause of industrial air pollution?
a. ozone plants
b. forest fires
c. car exhaust
d. burning fossil fuels

Indoor Air Pollution

______ **8.** What is one way to reduce indoor air pollution?
a. Use less fossil fuels.
b. Use more fossil fuels.
c. Use chemical solvents.
d. Improve ventilation.

ACID PRECIPITATION

______ **9.** What do high concentrations of acids in rain, sleet, or snow help to
cause?
a. ozone rain
b. acid precipitation
c. dirty snow
d. nitric imbalance

Acid Precipitation and Plants

______ **10.** Which of the following is an effect of acidification?
a. decreased soil acidity
b. increased plant nutrients
c. unchanged soil chemistry
d. increased soil acidity

Directed Reading A *continued*

The Effects of Acid Precipitation on Forests

______ **11.** What has acid precipitation done to some forests?
- **a.** It has made them stronger.
- **b.** It has damaged large areas.
- **c.** It hasn't done anything.
- **d.** It has made them grow.

Acid Precipitation and Aquatic Ecosystems

______ **12.** What is a very quick change in acidity in a body of water called?
- **a.** basic shock
- **b.** toxic shock
- **c.** acid shock
- **d.** ozone shock

THE OZONE HOLE

______ **13.** What does the ozone hole enable to reach Earth's surface?
- **a.** acid rain
- **b.** primary pollutants
- **c.** convection currents
- **d.** ultraviolet radiation

Cooperation to Reduce the Ozone Hole

______ **14.** What kind of molecules made the ozone layer break down?
- **a.** oxygen
- **b.** BHB
- **c.** CFC
- **d.** nitrogen

AIR POLLUTION AND HUMAN HEALTH

______ **15.** What are some of the bad effects air pollution has on people?
- **a.** lung cancer, coughing
- **b.** sleepiness, baldness
- **c.** deafness, weight gain
- **d.** foot problems, weight loss

CLEANING UP AIR POLLUTION

______ **16.** What did the Clean Air Act give the EPA the power to do?
- **a.** control water pollution levels
- **b.** control air pollution levels
- **c.** control ultraviolet radiation
- **d.** control oxygen levels

| Directed Reading A *continued*

Controlling Air Pollution from Industry

______ **17.** What device is used to take out some pollutants from power plants?
   a. sponges
   b. scrapers
   c. vacuums
   d. scrubbers

The Allowance Trading System

______ **18.** What does the Allowance Trading System help companies to do?
   a. reduce pollution
   b. trade stock
   c. increase investments
   d. increase acidification

Reducing Air Pollution from Vehicles

______ **19.** What two kinds of power do hybrid cars use?
   a. water and gasoline
   b. gasoline and electricity
   c. electricity and oxygen
   d. steam and gasoline

Directed Reading B

Section: Characteristics of the Atmosphere

______ **1.** A mixture of gases surrounding a planet or moon is the
   a. oxygen.
   b. atmosphere.
   c. breathable air.
   d. hemisphere.

THE COMPOSITION OF THE ATMOSPHERE

______ **2.** The most common atmospheric gas on Earth is
   a. oxygen.
   b. argon.
   c. nitrogen.
   d. carbon dioxide.

______ **3.** Living things, such as plants, produce the atmosphere's
   a. oxygen.
   b. argon.
   c. nitrogen.
   d. carbon dioxide.

______ **4.** Most water in the atmosphere is in
   a. rain.
   b. ice.
   c. water vapor.
   d. carbon dioxide.

AIR PRESSURE AND TEMPERATURE

______ **5.** At sea level, a square inch of surface area is under almost how many
   pounds of pressure?
   a. 150 lb
   b. 15 lb
   c. 30 lb
   d. 1500 lb

______ **6.** Gas molecules in the atmosphere are pulled toward Earth by
   a. air pressure.
   b. the moon.
   c. gravity.
   d. surface area.

Directed Reading B *continued*

---

7. The measure of the force with which air molecules push on a surface

is called ___________________.

8. What happens to air pressure as you move away from Earth's surface?

____

____

9. Why are some parts of the atmosphere warmer than others?

____

____

LAYERS OF THE ATMOSPHERE

Match the correct definition with the correct term. Write the letter in the space provided.

_______ **10.** coldest layer of the atmosphere

_______ **11.** atmosphere layer including the ozone layer

_______ **12.** layer of atmosphere closest to Earth

_______ **13.** uppermost layer of the atmosphere

a. troposphere

b. mesosphere

c. stratospherc

d. thermosphere

14. How are the layers of the atmosphere defined?

____

____

15. In the troposphere and mesosphere, what happens to the temperature as altitude increases?

____

____

____

16. Why do the stratosphere and thermosphere have high temperatures?

____

____

____

Directed Reading B

Section: Atmospheric Heating

______ **1.** What is the travel time for solar energy to reach Earth?
   a. about 8 hours
   b. about 80 hours
   c. about 8 minutes
   d. about 8 days

RADIATION: ENERGY TRANSFER BY WAVES

______ **2.** What percentage of the energy radiated by the sun reaches
   Earth's surface?
   a. two-fiftieths
   b. two-thousandths
   c. two-millionths
   d. two-billionths

______ **3.** What percentage of the sun's energy that reaches Earth is absorbed by
   Earth's surface?
   a. 25%
   b. 50%
   c. 20%
   d. 5%

______ **4.** What percentage of the sun's energy that reaches Earth is absorbed by
   ozone, clouds, and atmospheric gases?
   a. 25%
   b. 50%
   c. 20%
   d. 5%

5. The collection of all frequencies or wavelengths of electromagnetic waves is

called ___________________________.

6. Solar radiation is ___________________________ by the different layers of Earth's

atmosphere.

7. Most solar energy that reaches Earth's surface takes the form of

___________________________.

Directed Reading B *continued*

CONDUCTION: ENERGY TRANSFER BY CONTACT

8. How is heat transferred in conduction?

____

____

____

9. Why is heat always transferred from warmer areas to colder areas?

____

____

____

____

CONVECTION: ENERGY TRANSFER BY MOTION

Match the correct description with the correct term. Write the letter in the space provided.

______ **10.** transfer of energy by circulation or movement of a liquid or gas

______ **11.** circular movement of warm fluid rising and cool fluid sinking

a. convection

b. convection current

THE GREENHOUSE EFFECT

12. Explain what process produces the greenhouse effect.

____

____

____

____

13. The balance between incoming solar energy and outgoing energy radiated

into space is called ___________________ and makes Earth livable.

❘ Directed Reading B *continued*

GLOBAL WARMING

14. A gradual increase in average global temperature is called

_________________________ .

15. What are greenhouse gases?

____

16. What human activities may increase the level of greenhouse gases in the atmosphere?

____

____

Directed Reading B

Section: Air Movement and Wind

WHAT CAUSES WIND?

_______ **1.** What causes differences in air pressure?
   a. even heating of Earth
   b. even cooling of Earth
   c. uneven heating of Earth
   d. increased heating of Earth

_______ **2.** The movement of air caused by differences in air pressure is called
   a. dense air.
   b. wind.
   c. polar air.
   d. vents.

_______ **3.** Air is warmer and less dense than surrounding air at the equator because the equator receives more
   a. wind.
   b. air pressure.
   c. solar energy.
   d. radiation.

_______ **4.** Because air at the poles is colder and denser than surrounding air, it
   a. rises.
   b. sinks.
   c. circulates.
   d. stagnates.

_______ **5.** High pressure areas are created around the poles as cold air
   a. rises.
   b. blows.
   c. stagnates.
   d. sinks.

6. When the paths of winds and ocean currents curve because of

Earth's rotation, it's called the _____________________.

| **Directed Reading B** *continued*

GLOBAL WINDS

Match the correct description with the correct term. Write the letter in the space provided.

_______ **7.** winds that blow from 30° latitude in both hemispheres almost to the equator

a. polar easterlies

b. westerlies

_______ **8.** winds formed when convection cells, pressure belts, and winds combine with the Coriolis effect

c. trade winds

d. global winds

_______ **9.** wind formed as cold, sinking air moves from the poles to 60° north and 60° south latitude

_______ **10.** wind belts that extend between 30° and 60° latitude in both hemispheres

LOCAL WINDS

11. How can the properties of matter making up Earth's surface cause local winds?

____

____

____

Match the correct description with the correct term. Write the letter in the space provided.

_______ **12.** cool air that flows over the ocean toward the land during the day

a. valley breeze

b. land breeze

_______ **13.** cool air that flows over the land toward the ocean during the night

c. sea breeze

d. mountain breeze

_______ **14.** warm air that rises up mountain slopes during the day

_______ **15.** cool air that moves down mountain slopes during the night

Directed Reading B

Section: The Air We Breathe

AIR POLLUTION

1. The contamination of the atmosphere by the introduction of pollutants from

   human and natural sources is _____________________.

2. Examples of primary air pollutants in nature are dust, sea salt, and

   _____________________.

3. How are secondary pollutants formed?

   ____

   ____

   ____

4. List two examples of secondary pollutants.

   ____

   ____

5. What is one reason that ozone near Earth's surface is dangerous?

   ____

   ____

6. How is smog formed?

   ____

   ____

   ____

   ____

7. What is one way local geography plays a part in smog formation in Los
   Angeles?

   ____

   ____

   ____

| Directed Reading B *continued*

HUMAN-CAUSED AIR POLLUTION

_______ **8.** How much of the human-caused air pollution in the United States is caused by cars?
   a. about 10% to 20%
   b. about 25% to 35%
   c. about 45% to 55%
   d. about 60% to 70%

9. List two sources of industrial air pollution.

____

____

____

10. What are two ways to reduce indoor air pollution?

____

____

ACID PRECIPITATION

_______ **11.** Acid precipitation includes rain, sleet, or snow with a high concentration of acids from
   a. water pollution.
   b. sulfuric acid.
   c. air pollution.
   d. nitric acid.

_______ **12.** When sulfur dioxide and nitrogen oxide combine with water in the atmosphere, they form
   a. sulfuric acid and carbon dioxide.
   b. sulfuric acid and nitric acid.
   c. nitric acid and carbon dioxide.
   d. nitric acid and citric acid.

_______ **13.** When acid precipitation causes the acidity of soil to increase, it is called
   a. calcification.
   b. acidification.
   c. percolation.
   d. deforestation.

Directed Reading B *continued*

14. Where are three of the forest areas in the world that are affected by acid precipitation located?

____

____

____

15. A rapid change in the acidity of a body of water is called

_______________________.

THE OZONE HOLE

16. What is the main problem caused by the ozone hole?

____

____

____

17. CFC molecules can remain active in the stratosphere for

_______________________.

18. What is one reason the ozone hole is dangerous to humans?

____

____

AIR POLLUTION AND HUMAN HEALTH

19. What are three effects of air pollution on human health?

____

____

____

CLEANING UP AIR POLLUTION

______ **20.** In the United States, the law that gives the Environmental Protection Agency (EPA) the authority to control the amount of air pollution is the
   a. Clean Environment Act.
   b. Cleaner World Act.
   c. Clean Air Act.
   d. Air Quality Act.

Directed Reading B *continued*

21. What are two methods industries use to reduce air pollution?

____

____

22. How is the Allowance Trading System used to reduce air pollution?

____

____

____

23. What are two ways car manufacturers are reducing air pollution?

____

____

24. What are two ways people can reduce pollution from vehicles?

____

____

Name _________________________________ Class _______________ Date _____________

Vocabulary and Section Summary A

Characteristics of the Atmosphere

VOCABULARY

In your own words, write a definition of the following terms in the space provided.

1. atmosphere

2. air pressure

3. troposphere

4. stratosphere

5. mesosphere

6. thermosphere

Vocabulary and Section Summary A *continued*

SECTION SUMMARY

Read the following section summary.

- Nitrogen and oxygen make up most of Earth's atmosphere.

- Air pressure decreases as altitude increases.

- The composition of atmospheric layers affects their temperature.

- The troposphere is the lowest atmospheric layer. It is the layer in which we live.

- The stratosphere contains the ozone layer, which protects us from harmful ultraviolet radiation.

- The mesosphere is the coldest atmospheric layer.

- The thermosphere is the uppermost layer of the atmosphere.

Skills Worksheet)

Vocabulary and Section Summary A

Atmospheric Heating

VOCABULARY

In your own words, write a definition of the following terms in the space provided.

1. radiation

2. electromagnetic spectrum

3. conduction

4. convection

5. convection current

6. greenhouse effect

Vocabulary and Section Summary A *continued*

SECTION SUMMARY

Read the following section summary.

- Energy travels from the sun to Earth by radiation. This energy drives many processes at Earth's surface.

- Energy in Earth's atmosphere is transferred by radiation, conduction, and convection.

- Radiation is the transfer of energy through space or matter by waves.

- Conduction is the transfer of energy by direct contact.

- Convection is energy transfer by the movement of matter.

Vocabulary and Section Summary A

Air Movement and Wind

VOCABULARY

In your own words, write a definition of the following terms in the space provided.

1. wind

____

____

____

2. Coriolis effect

____

____

____

SECTION SUMMARY

Read the following section summary.

- Winds blow from areas of high pressure to areas of low pressure.
- Pressure belts are caused by the uneven heating of Earth's surface by the sun.
- The Coriolis effect causes wind to appear to curve as it moves across Earth's surface.
- Global winds include the polar easterlies, the westerlies, and the trade winds.
- Local winds include sea and land breezes and valley and mountain breezes.

Vocabulary and Section Summary A

The Air We Breathe

VOCABULARY

In your own words, write a definition of the following terms in the space provided.

1. air pollution

____

____

____

2. acid precipitation

____

____

____

SECTION SUMMARY

Read the following section summary.

- Air pollution is the introduction of harmful substances into the air by humans or by natural events.

- Primary pollutants are pollutants that are put directly into the air by human or natural activity.

- Secondary pollutants are pollutants that form when primary pollutants react with other primary pollutants or with naturally occurring substances.

- Transportation, industry, and natural sources are the main sources of air pollution.

- The burning of fossil fuels may lead to air pollution and acid precipitation, which may harm human and wildlife habitats.

- Air pollution can be reduced by legislation, such as the Clean Air Act; by technology, such as scrubbers; and by changes in lifestyle.

Vocabulary and Section Summary B

Characteristics of the Atmosphere

VOCABULARY

After you finish reading the section, try this puzzle! Use the clues below to solve the crossword puzzle.

ACROSS

2. the coldest layer of the atmosphere

5. the layer of the atmosphere where we live

6. the uppermost atmospheric layer

DOWN

1. mixture of gases that surrounds Earth

3. atmospheric layer above the troposphere

4. the measure of the force with which air molecules are pushing on Earth's surface

| Vocabulary and Section Summary B *continued*

SECTION SUMMARY

Read the following section summary.

- Nitrogen and oxygen make up most of Earth's atmosphere.
- Air pressure decreases as altitude increases.
- The composition of atmospheric layers affects their temperature.
- The troposphere is the lowest atmospheric layer. It is the layer in which we live.
- The stratosphere contains the ozone layer, which protects us from harmful ultraviolet radiation.
- The mesosphere is the coldest atmospheric layer.
- The thermosphere is the uppermost layer of the atmosphere.

Vocabulary and Section Summary B

Atmospheric Heating
VOCABULARY

After you finish reading the section, try this puzzle! The underlined words are missing all their vowels. Write the completed words in the spaces provided.

1. the transfer of heat due to the movement of a liquid or gas: <u>CNVCTN</u>

____

2. all the frequencies or wavelengths of electromagnetic radiation:
<u>LCTRMGNTC SPCTRM</u>

____

3. the transfer of energy as heat through a material: <u>CNDCTN</u>

____

4. the transfer of energy as electromagnetic waves: <u>RDTN</u>

____

5. a circular movement of matter that results from differences in density:
<u>CNVCTN CRRNT</u>

____

6. the warming of Earth's surface and lower atmosphere that occurs when water vapor, carbon dioxide, and other gases absorb and reradiate thermal energy:
<u>GRNHS FFCT</u>

____

| Vocabulary and Section Summary B *continued*

SECTION SUMMARY

Read the following section summary.

- Energy travels from the sun to Earth by radiation. This energy drives many processes at Earth's surface.
- Energy in Earth's atmosphere is transferred by radiation, conduction, and convection.
- Radiation is the transfer of energy through space or matter by waves.
- Conduction is the transfer of energy by direct contact.
- Convection is energy transfer by the movement of matter.

Vocabulary and Section Summary B

Air Movement and Wind

VOCABULARY

After you finish reading the section, try this puzzle! In the space provided, write the term described. Then, find the words in the word search puzzle on the following page. Words are hidden vertically, horizontally, diagonally, and backward.

_________________________ **1.** the movement of air caused by differences in air pressure

_________________________ **2.** the curving of the path of a moving object from an otherwise straight path due to Earth's rotation

_________________________ **3.** a mixture of gases that surrounds a planet or moon

_________________________ **4.** large air-circulation patterns produced by a combination of convection cells, pressure belts, winds, and the Coriolis effect

_________________________ **5.** small air-circulation patterns that move short distances and can blow from any direction

_________________________ **6.** the transfer of energy as heat through a material

_________________________ **7.** the movement of matter due to differences in density and the transfer of energy that results from this movement

_________________________ **8.** the transfer of energy as electromagnetic waves

| Vocabulary and Section Summary B *continued*

Q	T	F	H	Y	Q	K	M	I	D	J	L	B	B	H	Y	D	L	C	L
O	A	J	Q	U	H	A	M	X	F	U	Y	Q	V	Q	T	J	O	N	O
I	N	U	I	W	Y	I	G	M	J	R	V	M	N	D	C	R	U	N	C
L	Z	T	K	Z	K	N	H	D	B	T	J	T	S	J	I	Z	O	U	A
H	Z	J	H	Z	H	M	E	U	R	D	V	D	F	O	J	I	D	U	L
R	T	U	H	Y	R	V	D	V	Z	E	I	R	L	R	T	P	K	W	W
R	X	S	T	H	S	N	P	U	H	Z	R	I	X	C	W	D	R	T	I
G	L	O	B	A	L	W	I	N	D	S	S	E	U	D	G	I	H	K	N
Z	H	R	B	W	H	S	Y	Q	E	E	V	D	H	M	K	T	N	R	D
K	V	H	W	X	D	L	P	E	F	A	N	R	I	P	R	V	K	D	S
B	M	N	Y	V	X	Q	W	E	C	O	P	I	V	Y	S	O	J	G	I
S	Y	I	U	O	W	O	C	G	C	S	B	B	Z	M	N	O	J	F	L
B	B	M	E	Q	E	T	R	Y	I	G	B	F	N	U	K	O	M	J	V
X	R	Q	R	V	S	P	Y	T	V	M	Q	B	F	V	O	K	U	T	K
R	A	D	I	A	T	I	O	N	G	S	Y	X	S	G	W	T	R	A	A
C	B	U	E	W	T	Z	X	R	L	Z	B	C	W	V	M	H	H	A	R
N	E	M	A	N	F	P	L	L	J	Q	B	O	M	L	M	W	W	H	A
J	K	Z	G	Z	W	H	Y	L	C	P	Z	S	J	T	V	G	R	H	C
E	H	S	F	S	Y	F	U	N	O	I	T	C	E	V	N	O	C	D	L

SECTION SUMMARY

Read the following section summary.

- Winds blow from areas of high pressure to areas of low pressure.
- Pressure belts are caused by the uneven heating of Earth's surface by the sun.
- The Coriolis effect causes wind to appear to curve as it moves across Earth's surface.
- Global winds include the polar easterlies, the westerlies, and the trade winds.
- Local winds include sea and land breezes and valley and mountain breezes.

Vocabulary and Section Summary B

The Air We Breathe

VOCABULARY

After you finish reading the section, try this puzzle! Look at the clues below, and write the terms being described in the blanks provided. Then, write the boxed letters in the space provided to spell out a phrase related to the atmosphere.

1. a chemical that causes ozone to break down into oxygen, which does not block harmful UV rays

2. pollutants that are put directly into the air by human or natural activity

3. a device that is used to remove some pollutants before they are released by smokestacks

4. the contamination of the atmosphere by the introduction of pollutants from human and natural sources

5. rain, sleet, or snow that contains a high concentration of acids

6. the mixture of indoor air with outdoor air

7. the increase of soil acidity due to acid precipitation

8. pollutants that form when primary pollutants react with other primary pollutants or with naturally occurring substances, such as water vapor

9. a rapid change in the acidity of a body of water

1. __ __ ☐

2. __ __ __ __ __ __ __   __ __ ☐ __ __ __ __ __ __ __

3. __ __ __ __ __ __ ☐ __

4. ☐ __ __   __ __ __ __ __ __ __ __ __ __

5. __ __ __ __   __ __ __ __ __ __ __ __ __ __ __ __ __ ☐

6. __ ☐ __ __ __ __ __ __ __ __ __ __

7. ☐ __ __ __ __ __ __ __ __ __ __ __ __ __

8. __ __ __ __ __ __ __ ☐ __   __ __ __ __ __ __ __ __ __ ☐ __

9. __ __ __ __ __ ☐ __ __ __

10. What is the phrase?

____

Vocabulary and Section Summary B *continued*

SECTION SUMMARY

Read the following section summary.

- Air pollution is the introduction of harmful substances into the air by humans or by natural events.

- Primary pollutants are pollutants that are put directly into the air by human or natural activity.

- Secondary pollutants are pollutants that form when primary pollutants react with other primary pollutants or with naturally occurring substances.

- Transportation, industry, and natural sources are the main sources of air pollution.

- The burning of fossil fuels may lead to air pollution and acid precipitation, which may harm human and wildlife habitats.

- Air pollution can be reduced by legislation, such as the Clean Air Act; by technology, such as scrubbers; and by changes in lifestyle.

Reinforcement

Earth's Amazing Atmosphere

Complete this worksheet after you finish reading the section "Characteristics of the Atmosphere."

Earth's atmosphere is divided into four layers. Choose the layer in Column B that best matches the description in Column A, and write your answer in the space provided. Then, use the directions below to label the diagram of Earth's atmosphere on the next page.

Column A

______ **1.** the layer of Earth's atmosphere you live in

______ **2.** the coldest layer of Earth's atmosphere; lies directly below the uppermost layer

______ **3.** the uppermost layer of the atmosphere

______ **4.** the layer that contains most of the atmosphere's ozone; above the layer that you live in

Column B

a. troposphere

b. stratosphere

c. mesosphere

d. thermosphere

5. Label the four layers of the atmosphere on the diagram on the next page.

6. There is no clear boundary between the uppermost layer of the atmosphere and space. The atmosphere becomes thinner and thinner and blends into space. At the very top of the diagram, write the word space with an arrow pointing up.

7. The ozone layer is the upper part of the atmospheric layer that contains most of the atmosphere's ozone. Use the symbol for ozone to draw in the ozone layer on the diagram.

8. The ozone layer is important because it absorbs ultraviolet radiation. Draw a wavy line coming from space to represent the UV radiation that is absorbed by the ozone layer.

9. The troposphere is the densest layer of the atmosphere. It is much denser than the other layers. Shade this layer heavily to indicate how dense it is.

10. The stratosphere is very thin. Shade this layer lightly.

11. The mesosphere is even less dense than the stratosphere. Shade this layer very lightly.

Reinforcement *continued*

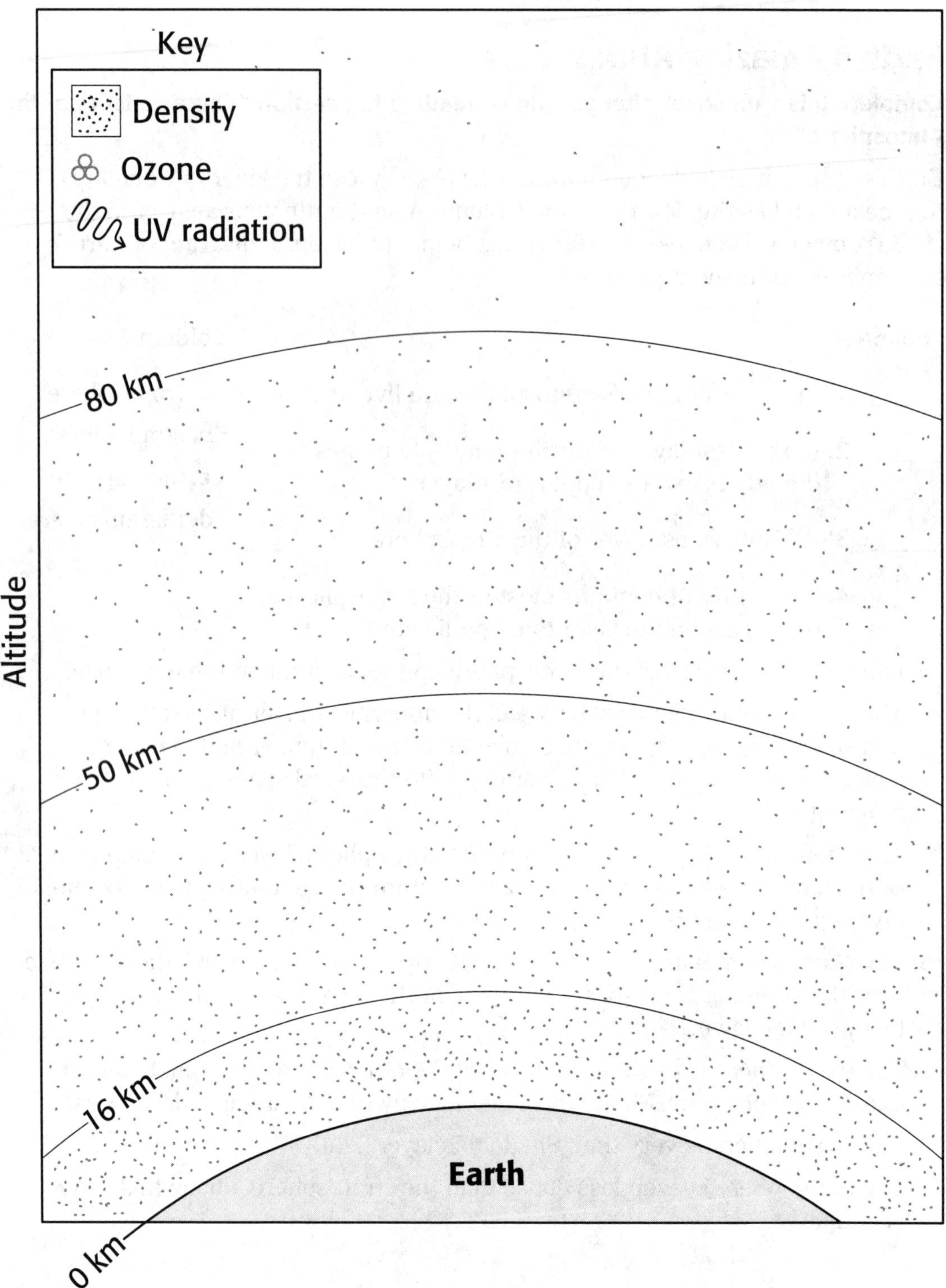

Critical Thinking

The Extraordinary GBG5K

While standing in line at the local Sack-n-Spend, you spot a tabloid newspaper. Next to the headline "Man Gives Birth to Baby Seal," you read this article:

SCIENTIST FINDS AIR POLLUTION SOLUTION

Bolognaville, Antarctica: Charlie Bloomfritter, "Earth scientist extraordinaire," has invented a sure-fire way to end air pollution.

Says Mr. Bloomfritter, "With my patented Gunk-B-Gone 5000 filtration machine, we can stop worrying about air pollution. As you can see, all solid particles in the air that go into the GBG5K do not come out. They are trapped inside forever. I have plans to build a super-large version to place in the parking lot outside, which will be so powerful that it will filter all the gunk out of Earth's atmosphere. People can now pollute all they want, because the GBG5K will clean up the mess."

Bloomfritter is a self-educated "gunkologist." He wrote the article "Building a Better Time Machine" in last week's issue.

EVALUATING SOURCES

1. Does this publication seem like a reliable source of scientific facts? Explain.

____

____

____

____

READING FOR DETAIL

2. What kinds of pollutants does Mr. Bloomfritter claim the GBG5K will catch?

____

____

____

____

____

| Critical Thinking *continued*

DEMONSTRATING REASONED JUDGMENT

3. Mr. Bloomfritter says that the pollution particles will be trapped inside the machine "forever." What is wrong with this claim?

____

____

____

____

IDENTIFYING ASSUMPTIONS

4. List two incorrect assumptions Mr. Bloomfritter makes about air pollution.

____

____

____

____

SUGGESTING ALTERNATIVES

5. What changes would you recommend to Mr. Bloomfritter that would make his filtration system more effective against air pollution?

____

____

____

____

SciLinks Activity

OZONE: FRIEND OR FOE?

Go to www.scilinks.org. To find links related to the composition of the atmosphere, type in the keyword HY70328. Then, use the links to answer the following questions about ozone.

Internet Resources

For a variety of links related to this chapter, go to www.scilinks.org

Topic: Composition of the Atmosphere

SciLinks code: HY70328

1. What is ozone? Is it good or bad? Explain.

2. How does ozone affect human health?

3. How can you tell if you are being affected by ozone?

4. What can you do to reduce ozone levels?

Section Review

Characteristics of the Atmosphere

UNDERSTANDING CONCEPTS

1. Applying Why does the temperature of different layers of the atmosphere vary?

____

____

____

2. Identifying What is air pressure?

____

____

3. Summarizing Why does air pressure decrease as altitude increases?

____

____

____

INTERPRETING GRAPHICS

Use the graph below to answer the next two questions.

Composition of the Atmosphere

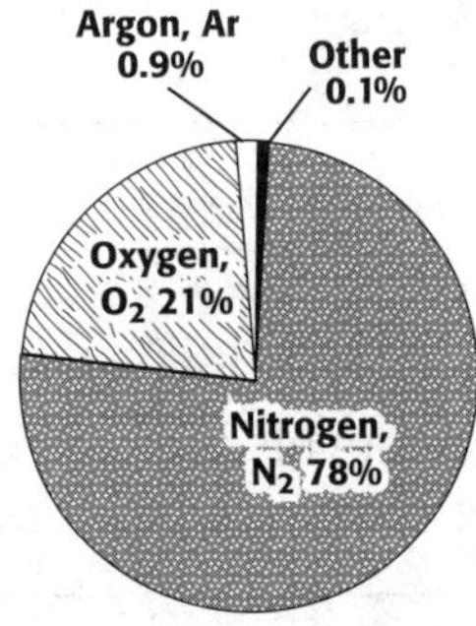

4. Identifying What two gases make up most of the atmosphere? What is the combined percentage of the two gases?

____

____

5. Identifying What percentage of the atmosphere is argon?

____

Section Review *continued*

CRITICAL THINKING

6. Applying Concepts Apply what you know about the relationship between altitude and air pressure to explain why rescue helicopters have a difficult time flying at altitudes above 6,000 m.

____

____

____

____

____

7. Making Comparisons Compare the four atmospheric layers. How do temperatures and pressures vary between layers?

____

____

____

____

____

MATH SKILLS

8. Making Calculations The troposphere ends at an altitude of about 12 km. The mesosphere begins at an altitude of about 50 km. What is the approximate thickness of the stratosphere? Show your work below.

Section Review

Atmospheric Heating

USING VOCABULARY

1. Use *conduction*, *radiation*, *convection*, and *global warming* in separate sentences.

____

____

____

____

UNDERSTANDING CONCEPTS

2. Applying How does the sun affect processes at Earth's surface?

____

____

3. Describing What is radiation?

____

____

____

4. Identifying Identify three ways that energy is transferred in the atmosphere.

____

____

5. Describing How does radiation differ from conduction and convection?

____

____

____

6. Summarizing Describe the electromagnetic spectrum, and explain how this energy reaches Earth.

____

____

____

CRITICAL THINKING

7. Making Comparisons How does conduction differ from convection?

____

____

____

____

8. Identifying Relationships How does the process of convection rely on conduction?

____

____

____

____

9. Making Comparisons What is the difference between the greenhouse effect and global warming?

____

____

____

____

| Section Review *continued*

MATH SKILLS

10. Making Calculations Find the average of the following temperatures: 22.8°C, 21.7°C, 12.5°C, 18.6°C, 25.7°C, 27.7°C, and 27.8°C. Show your work below.

CHALLENGE

11. Predicting Consequences How do you think processes of conduction and convection on Earth would be different if radiation from the sun were not constant or uniform?

____

____

____

____

____

____

____

____

____

____

____

____

____

Section Review

Air Movement and Wind

USING VOCABULARY

1. Write an original definition for *wind* and *Coriolis effect.*

UNDERSTANDING CONCEPTS

2. Applying Why do pressure belts form in the atmosphere?

3. Summarizing What causes winds?

4. Identifying What causes the Coriolis effect?

5. Applying How does the Coriolis effect affect wind movement?

6. Comparing How do the processes that cause global winds and local winds differ?

Section Review *continued*

---

7. Summarizing Describe the difference between sea and land breezes and between valley and mountain breezes.

____

____

____

CRITICAL THINKING

8. Analyzing Ideas Would there be winds if Earth's surface were the same temperature everywhere? Explain your answer.

____

____

____

____

9. Applying Concepts Imagine that you are near an ocean in the daytime. How could a local wind help you find the ocean if you don't know the way there?

____

____

____

____

CHALLENGE

10. Forming Hypotheses In the Northern Hemisphere, why do westerlies flow from the west but trade winds flow from the east?

____

____

____

____

Name _________________________________ Class _______________ Date ___________

Section Review

The Air We Breathe

USING VOCABULARY

Correct each statement by replacing the underlined term.

1. <u>Air pollution</u> is a sudden change in the acidity of a stream or lake.

   ____

2. <u>Smog</u> is rain, sleet, or snow that has a high concentration of acid.

   ____

UNDERSTANDING CONCEPTS

3. **Summarizing** How does smog form?

   ____

   ____

   ____

   ____

4. **Comparing** What is the difference between primary and secondary pollutants?

   ____

   ____

   ____

   ____

5. **Identifying** List five sources of indoor air pollution.

   ____

   ____

   ____

   ____

6. **Applying** How are fossil fuels related to air pollution and acid precipitation?

   ____

   ____

   ____

| Section Review *continued*

---

7. Listing List five short-term effects of air pollution on human health.

____

____

____

CRITICAL THINKING

8. Analyzing Ideas Is all air pollution caused by humans? Explain your answer.

____

____

____

____

____

9. Expressing Opinions How do you think that nations should resolve air-pollution problems that cross national boundaries?

____

____

____

____

____

10. Making Inferences Why may establishing a direct link between air pollution and health problems be difficult?

____

____

____

____

____

| Section Review *continued*

INTERPRETING GRAPHICS

The map below shows the pH of precipitation measured at field stations in the northeastern United States. On the pH scale, the lower a number is, the more acidic the solution is.

Use the map below to answer the next two questions.

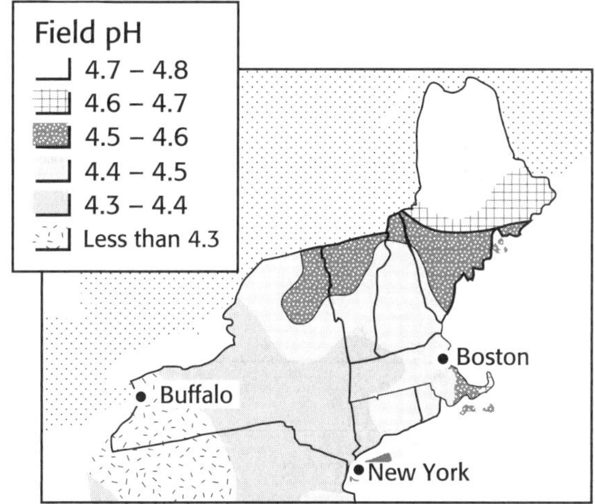

11. Forming Hypotheses Which areas have the most acidic precipitation? Hypothesize why.

____

____

____

12. Making Inferences Boston is a larger city than Buffalo is, but the precipitation measured in Buffalo is more acidic than the precipitation in Boston is. Explain why.

____

____

____

CHALLENGE

13. Making Inferences Do you think that air pollution can extend beyond its source area? How far? How would the pollution move?

____

____

____

Skills Worksheet

Chapter Review

USING VOCABULARY

1. Academic Vocabulary Which term—*transfer* or *result in*—best completes the

following sentence: Convection currents ______________________ heat in

the atmosphere.

For each pair of terms, explain how the meanings of the terms differ.

2. *conduction* and *convection*

____

____

____

3. *radiation* and *electromagnetic spectrum*

____

____

____

4. *wind* and *air pressure*

____

____

____

UNDERSTANDING CONCEPTS

Multiple Choice

_______ **5.** What is the most abundant gas in the atmosphere?
- **a.** oxygen
- **b.** hydrogen
- **c.** nitrogen
- **d.** carbon dioxide

_______ **6.** The ozone layer is located in the
- **a.** stratosphere.
- **b.** troposphere.
- **c.** thermosphere.
- **d.** mesosphere.

_______ **7.** By which method does most thermal energy in the atmosphere
circulate?
   a. conduction
   b. convection
   c. advection
   d. radiation

Short Answer

8. Classifying Classify the following pollutants as primary or secondary
pollutants: car exhaust, acid precipitation, smoke from a factory, and fumes
from burning plastic.

____

____

9. Analyzing Why does air pressure decrease as altitude increases?

____

____

10. Describing Explain why air rises when it is heated.

____

____

11. Identifying What is the main cause of temperature changes in the
atmosphere?

____

____

12. Summarizing What are secondary pollutants, and how do they form? Give an
example of a secondary pollutant.

____

____

____

13. Summarizing What is the electromagnetic spectrum? How much of the
electromagnetic spectrum is visible light?

____

____

____

| Chapter Review *continued*

---

14. Describing Describe the relationship between solar energy and global winds.

____

____

____

INTERPRETING GRAPHICS

Use the pie graph below to answer the next two questions.

Total World Emissions of CO_2

15. Identifying Which produced more CO_2 emissions in 1995: developing countries or developed countries?

____

____

____

16. Listing Which developed country or group of countries had the highest CO_2 emissions?

____

____

____

____

▌Chapter Review *continued*

WRITING SKILLS

17. Writing Persuasively Write a letter to a family member or friend. Tell him or her about the effects of pollution and explain what he or she can do to reduce pollution.

____

____

____

____

____

____

____

____

____

____

18. Outlining Topics Outline the steps that you would take to reduce indoor air pollution at your school.

____

____

____

____

____

____

____

____

____

CRITICAL THINKING

19. Concept Mapping Use the following terms to create a concept map:
mesosphere, stratosphere, layers, temperature, troposphere, and *atmosphere.*

20. Identifying Relationships What is the relationship between the greenhouse effect and global warming?

____

____

____

____

| Chapter Review *continued*

---

21. Applying Concepts How do you think the Coriolis effect would change if Earth rotated twice as fast as it does? Explain your answer.

____

____

____

____

22. Identifying Relationships How are global winds related to convection?

____

____

____

____

23. Making Comparisons Compare local winds and global winds. Do the two types of winds form in the same way? Explain your answer.

____

____

____

____

24. Applying Concepts How is radiation fundamentally different from conduction and convection?

____

____

____

____

| Chapter Review *continued*

25. Making Inferences Is air a renewable or nonrenewable resource? Is stratospheric ozone a renewable or nonrenewable resource? Explain your answers.

____

____

____

____

____

____

26. Analyzing Processes How is heat transferred from the sun to Earth? What carries this energy from the sun to Earth?

____

____

____

____

____

____

27. Applying Concepts Compare the characteristics of visible light to the portions of the electromagnetic spectrum that are not visible. Which characteristics are different?

____

____

____

____

____

____

INTERPRETING GRAPHICS

Use the graph below to answer the next three questions.

**Concentration of Carbon Dioxide
in the Atmosphere, 1970–2000**

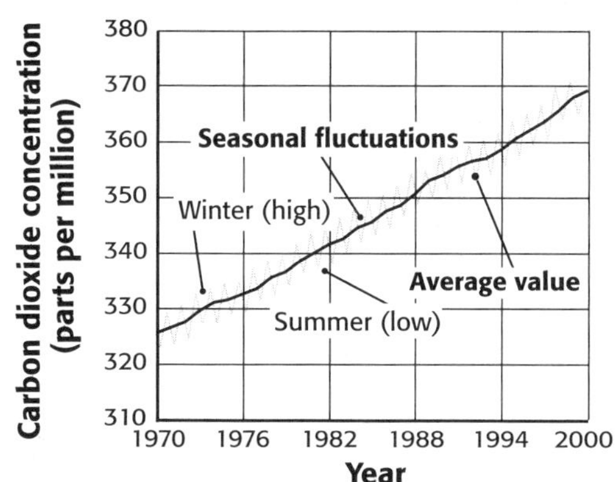

28. Evaluating Data According to the graph, is the carbon dioxide concentration increasing or decreasing with time?

____

____

29. Evaluating Data What factors could be contributing to the trend you see in the graph?

____

____

____

____

30. Making Inferences What can you infer about what will happen to the atmospheric carbon dioxide concentration in future years not shown on this graph? How could this result be avoided?

____

____

____

____

____

____

| Chapter Review *continued*

MATH SKILLS

31. Making Calculations Maximum local wind speeds for each of the last seven days were 12 km/h, 20 km/h, 11 km/h, 6 km/h, 8km/h, 19 km/h, and 17 km/h. What was the average maximum wind speed? Show your work below.

CHALLENGE

32. Predicting Consequences Imagine that there is a planet that has two suns. Assume that the planet and Earth have similar land and water conditions. Describe how two energy sources combined might affect the imaginary planet's winds.

____

____

____

____

____

____

____

____

____

____

____

____

____

Assessment

Chapter Pretest

Teacher Notes and Answer Key

The Pretest questions are designed to help you determine the prior knowledge of your students. Some questions test whether students have mastered the background knowledge they need to understand the content you are about to teach. Other questions test your students' prior knowledge of the content you are about to teach. Use the Pretest with the Test Doctors and diagnostic teaching tips in these notes pages to help you tailor your instruction to your students' specific needs.

QUESTION NUMBER	CORRECT ANSWER	STANDARD
1	B	6.3.a
2	B	6.3.c
3	D	6.4.a
4	A	6.4.d
5	D	6.3.c
6	B	6.3.d
7	C	6.4.b
8	A	6.4.d
9	D	6.6.a

TEST DOCTOR

The following Pretest questions have been diagnosed by the Test Doctor. Find out what might be causing your students' "ailing" answers. Each Test Doctor is followed by a diagnostic teaching tip to help you address students' learning needs.

Question 1 *asks students in what way water, light, and sound are alike.*

A **Incorrect.** Water and sound do not carry energy through radiation the way light does.

B **Correct.** Water, light, and sound all carry energy from one place to another in waves.

C **Incorrect.** Water, light, and sound are carried through waves, not rays.

D **Incorrect.** Convection occurs in water, but not in light or sound.

Diagnostic Teaching Tip: Students who have difficulty answering this question should benefit from a review of waves. Ask students to visualize energy movement as you create waves in a pan of water. Tell students that light and sound also travel in waves, although we cannot see these waves with our eyes. Discuss the movement of energy as waves with the class.

Question 2 *asks students to identify the choice that is the transfer of energy as the result of a movement of matter.*

A **Incorrect.** Condensation, or water vapor changing to liquid, is part of the water cycle.

B **Correct.** Convection is a transfer of energy as a result of the movement of matter.

Chapter Pretest *continued*

C Incorrect. Conduction does not involve a flow of matter.

D Incorrect. This answer refers to the Coriolis effect, the curving of wind and water from a straight path due to Earth's rotation.

Diagnostic Teaching Tip: Students who answer this question incorrectly might benefit from a review of convection and convection currents. Have students look up the word convection. Ask students what matter flows in an ocean convection current.

Question 3 *asks students how the sun powers winds that form surface currents in the ocean.*

A Incorrect. This is not the mechanism by which the sun powers winds that form surface currents.

B Incorrect. This is a far-fetched explanation that reveals a lack of knowledge about how the sun affects Earth's surface.

C Incorrect. Solar energy does not blow into Earth's atmosphere or displace air.

D Correct. The pressure differences between the equator and other latitudes cause winds that form the surface currents of the oceans.

Diagnostic Teaching Tip: Students who have difficulty with this question might benefit from a review of the relationship between the sun, air temperature, air pressure, and winds. Have small groups use a heat lamp to heat a pan of water and measure temperature changes. Ask students to write an explanation of how the sun transfers energy to Earth's surface. Then, discuss with the class how pressure differences cause winds.

Question 4 *asks students to identify the statement about convection currents in the ocean that is true.*

A Correct. This statement is true. Water is warm at the surface and is carried by surface currents to the polar regions where it cools, becomes more dense, and sinks to the ocean floor.

B Incorrect. Water is cold deep below the surface and moves in convection currents there. This water rises to replace surface water that leaves in surface currents. This answer is also wrong because the water does not cool when it gets to the tropics.

C Incorrect. Surface water is warm when it is carried to the polar regions. Surface water doesn't rise.

D Incorrect. Warm surface water doesn't travel to the tropics in ocean convection currents, nor does it sink.

Diagnostic Teaching Tip: Students who answer this question incorrectly might benefit from a review of ocean convection currents. Use a globe to show the movement of heat energy from warm regions to polar regions, and describe the process. Then, have students make a diagram to show the process.

Question 5 *asks students to demonstrate an understanding of what happens when cool air comes into direct contact with the warm surface of Earth.*

Chapter Pretest *continued*

A Incorrect. Convection is the transfer of energy through the movement of matter in a liquid or gas. When cool air comes in direct contact with the warm surface of Earth, heat is not transferred through the movement of matter.

B Incorrect. Cold air doesn't rise into the atmosphere.

C Incorrect. Heat is transferred from warm areas to cool areas.

D Correct. When cool air comes into direct contact with the warm surface of Earth, heat is transferred from the warmer area (Earth's surface) to the cooler area (the atmosphere) by conduction.

Diagnostic Teaching Tip: Students who have difficulty answering this question correctly might benefit from a demonstration of conduction. Have a student hold a hand near a light bulb, and ask the class to explain how heat energy is being transferred from the bulb to the air directly in contact with it. Be sure to emphasize Section 2, "Atmospheric Heating," in Chapter 14, "The Atmosphere." Students must know that heat flows in fluids by conduction in order to master standard 6.3.c.

Question 6 *asks students to identify the transfer of energy as electromagnetic waves between the sun and the surface of Earth.*

A Incorrect. Convection is the transfer of energy through the movement of matter in a liquid or gas. This is not the method in which energy is transferred as electromagnetic waves between the sun and Earth.

B Correct. The transfer of energy as electromagnetic waves between the sun and Earth occurs through radiation.

C Incorrect. This term may be confused with radiation.

D Incorrect. Conduction is the transfer of energy as heat by direct contact, not through electromagnetic waves.

Diagnostic Teaching Tip: Students who have difficulty answering this question correctly might benefit from a discussion of radiation, conduction, and convection. Have students work in small groups to list the processes on Earth that are driven by energy provided by the sun (for example, winds, the water cycle, ocean currents, changes in the weather). Be sure to emphasize Section 2, "Atmospheric Heating," in Chapter 14, "The Atmosphere." Students must know that heat energy is also transferred between objects by radiation, which can travel through space, in order to master standard 6.3.d.

Question 7 *asks students why most of the solar energy that reaches Earth is visible light.*

A Incorrect. Temperature is unrelated to forms of solar energy that reach Earth.

B Incorrect. Most visible light is not reflected by the atmosphere.

C Correct. The atmosphere absorbs almost all radiation with wavelengths shorter than those of visible light but absorbs only a small amount of visible light.

D Incorrect. Gammas rays are part of the electromagnetic spectrum, but are not visible light.

| Chapter Pretest *continued*

Diagnostic Teaching Tip: Students who have difficulty with this question might benefit from examining the effects of Earth's atmosphere on solar radiation, including the greenhouse effect. Have students work with a partner to research these effects. Have pairs write a short paragraph or make a diagram illustrating what they learned. Be sure to emphasize Section 2, "Atmospheric Heating," in Chapter 14, "The Atmosphere." Students must know that solar energy reaches Earth through radiation, mostly in the form of visible light, in order to master standard 6.4.b.

Question 8 *asks students to identify the process of heat being distributed in Earth's atmosphere as shown in the illustration.*

A **Correct.** This illustration of convection depicts warm air rising in the atmosphere and cold air sinking. The warm air rises because it is less dense than the cold air, which sinks. This movement of matter distributes the heat.

B **Incorrect.** Radiation is the transfer of energy through space or matter as electromagnetic waves.

C **Incorrect.** Conduction is the transfer of heat energy through direct physical contact.

D **Incorrect.** Condensation takes place when water vapor changes to liquid.

Diagnostic Teaching Tip: Students who have difficulty answering this question correctly might benefit from a discussion of convection and convection currents. Have students think of when they stood in front of a freezer door and opened it. Ask them what they observed and how they can explain it. Tell them that heat is transferred from warm areas to cooler areas. Be sure to emphasize Section 2, "Atmospheric Heating," in Chapter 14, "The Atmosphere." Students must be able to understand how convection currents distribute heat in the atmosphere in order to master standard 6.4.d.

Question 9 *asks students to identify which secondary pollutant is the result of the use of fossil fuels.*

A **Incorrect.** Exhaust is a primary pollutant.

B **Incorrect.** Smoke is a primary pollutant.

C **Incorrect.** Carbon monoxide is a primary pollutant.

D **Correct.** Ozone is a secondary pollutant that is formed when sunlight reacts with exhaust and air.

Diagnostic Teaching Tip: Students who have difficulty with this question might benefit from an activity examining pollutants. Ask the class to give examples of primary pollutants, or pollutants that are put directly into the air. Examples include dust, sea salt, volcanic gases and ash, smoke, carbon monoxide, and chemicals. Then, tell students that secondary pollutants, such as ozone and smog, are formed when primary pollutants react with each other or with other substances. Be sure to emphasize Section 4, "The Air We Breathe," in Chapter 14, "The Atmosphere." Students must know that the utility of energy sources is determined in part by the consequences of their use in order to master standard 6.6.a.

Name _________________________________ Class _______________ Date _______________

Chapter Pretest

______ **1.** In what way are water, light, and sound alike?
  A They all carry energy through radiation.
  B They all carry energy through waves.
  C They all carry energy through rays.
  D They all carry energy through convection.

______ **2.** The transfer of energy as the result of a movement of matter is called
  A condensation.
  B convection.
  C conduction.
  D Coriolis.

______ **3.** How does the sun power winds that form surface currents
  in the ocean?
  A The sun heats the atmosphere at the poles, forcing arctic air toward
    the tropics, and forming ocean currents.
  B Heat from the sun causes air to rise above the atmosphere, and
    when it sinks, winds form.
  C Solar energy from the sun blows into Earth's atmosphere, displaces
    air, and causes winds.
  D The sun heats the atmosphere more at the equator than other
    latitudes, causing pressure differences and winds.

______ **4.** Which statement is true of convection currents in the oceans?
  A Warm surface water is carried to polar regions, cools, and sinks.
  B Cold surface water is carried to the tropics, cools, and rises.
  C Cold surface water is carried to the polar regions, cools, and rises.
  D Warm surface water is carried to the tropics, warms, and sinks.

______ **5.** What happens to cool air when it comes into direct contact with the
  warm surface of Earth?
  A Heat is transferred to the atmosphere by convection.
  B Cold air rises into the atmosphere.
  C Cool air molecules transfer energy to Earth's warm surface.
  D Heat is transferred to the atmosphere by conduction.

______ **6.** Which word describes the transfer of energy as electromagnetic waves
  between the sun and the surface of Earth?
  A convection
  B radiation
  C radioactivity
  D conduction

_______ **7.** Most of the solar energy that reaches Earth is visible light. Why is this true?

    A Earth's temperature keeps other forms of solar energy from reaching Earth.

    B Most of the visible light is reflected by the atmosphere.

    C Only a small amount of visible light is absorbed by the atmosphere.

    D Visible light is composed of gamma rays.

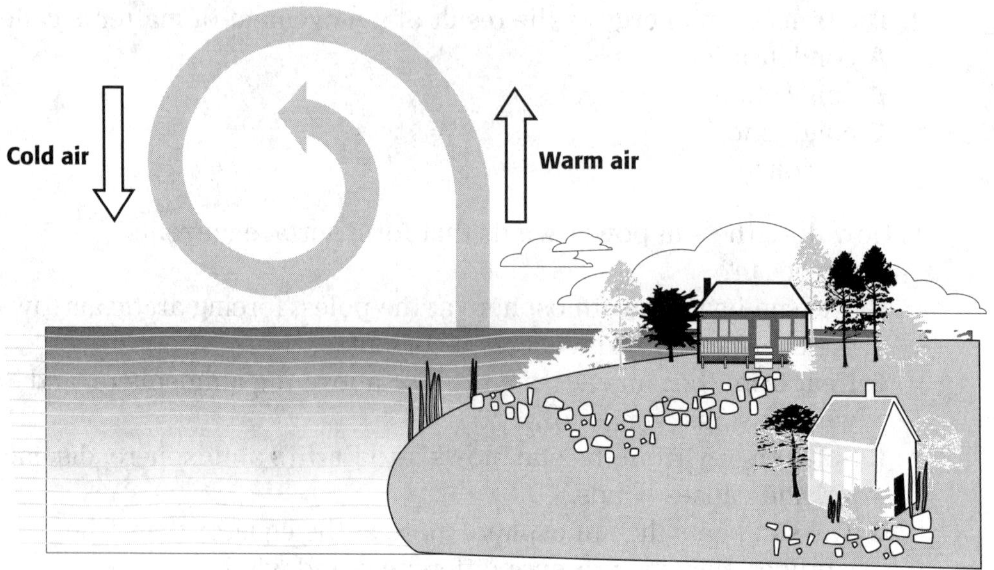

_______ **8.** The illustration above shows heat being distributed in Earth's atmosphere. What process is shown?

    A convection

    B radiation

    C conduction

    D condensation

_______ **9.** The burning of fossil fuels as an energy source results in the formation of which secondary pollutant?

    A exhaust

    B smoke

    C carbon monoxide

    D ozone

Section Quiz

Section: Characteristics of the Atmosphere

Write the letter of the correct answer in the space provided.

_______ **1.** Which answer best describes Earth's atmosphere?
    a. mostly oxygen and 21% nitrogen
    b. mostly nitrogen and 21% oxygen
    c. mostly carbon dioxide and 21% oxygen
    d. mostly nitrogen and 21% carbon dioxide

_______ **2.** What mainly causes differences in air temperature at different altitudes?
    a. Air radiates solar energy.
    b. Gases radiate light.
    c. Gases absorb solar energy.
    d. Moisture reflects sunlight.

_______ **3.** What are the two highest layers of the atmosphere?
    a. troposphere, mesosphere
    b. thermosphere, mesosphere
    c. stratosphere, thermosphere
    d. troposphere, stratosphere

_______ **4.** In what layer of the atmosphere is harmful UV radiation absorbed?
    a. mesosphere
    b. troposphere
    c. hemisphere
    d. stratosphere

Match the correct definition with the correct term. Write the letter in the space provided.

_______ **5.** layered gases, thin air, little moisture

_______ **6.** coldest layer, temperature decreases as altitude increases

_______ **7.** lack of particle density, little thermal energy transfer

_______ **8.** densest layer, contains almost 90% of the atmosphere's mass

    a. troposphere
    b. thermosphere
    c. stratosphere
    d. mesosphere

Name _________________________________ Class _______________ Date _____________

Section Quiz

Section: Atmospheric Heating

Write the letter of the correct answer in the space provided.

_______ **1.** How much energy radiated by the sun reaches Earth?
- **a.** about 80%
- **b.** about 50%
- **c.** about two-billionths
- **d.** about one one-hundredth

_______ **2.** Energy transferred as electromagnetic waves is called
- **a.** conduction.
- **b.** radiation.
- **c.** convection.
- **d.** convection current.

_______ **3.** Energy transferred as heat through a material is called
- **a.** conduction.
- **b.** radiation.
- **c.** convection.
- **d.** convection current.

_______ **4.** Thermal energy transferred by circulation of a liquid or gas is called
- **a.** conduction.
- **b.** radiation.
- **c.** convection.
- **d.** convection current.

_______ **5.** The process by which gases in the atmosphere absorb solar energy and reradiate it back to Earth is called
- **a.** the thermal effect.
- **b.** the greenhouse effect.
- **c.** global warming.
- **d.** radiation balance.

_______ **6.** When the amount of energy received from the sun and the amount of energy returned to space are about equal, it is called
- **a.** seasonal equality.
- **b.** radiation balance.
- **c.** solar reradiation.
- **d.** global warming.

_______ **7.** One reason for global warming may be
- **a.** decreasing global gases.
- **b.** increasing global gases.
- **c.** increasing greenhouse gases.
- **d.** decreasing greenhouse gases.

Section Quiz

Section: Air Movement and Wind

Write the letter of the correct answer in the space provided.

_______ **1.** What causes wind?
- **a.** differences in air pressure
- **b.** differences in gravity
- **c.** differences in oxygen
- **d.** differences in the thermosphere

_______ **2.** What causes differences in air pressure around Earth?
- **a.** Warm air rises at the equator, and cold air sinks at the poles.
- **b.** Warm air sinks at the equator, and cold air rises at the poles.
- **c.** Warm air rises at the equator, and cold air rises at the poles.
- **d.** Cold air rises at the equator, and warm air sinks at the poles.

_______ **3.** Air moves in patterns of high and low pressure that are part of
- **a.** pressure belts.
- **b.** convection currents.
- **c.** convection cells.
- **d.** trade winds.

_______ **4.** In the Northern Hemisphere, winds traveling north curve to the east because of
- **a.** trade winds.
- **b.** convection currents.
- **c.** the Coriolis effect.
- **d.** polar easterlies.

_______ **5.** Global winds that blow from west to east are called
- **a.** polar easterlies.
- **b.** westerlies.
- **c.** mountain breezes.
- **d.** trade winds.

_______ **6.** Global winds that blow almost to the equator from 30° latitude are called
- **a.** northerlies.
- **b.** trade winds.
- **c.** polar easterlies.
- **d.** global easterlies.

_______ **7.** A wind belt that carries cold arctic air over the United States is called a(n)
- **a.** polar easterly.
- **b.** jet stream.
- **c.** convection current.
- **d.** convection stream.

_______ **8.** Local winds are affected by
- **a.** global winds.
- **b.** global geographic features.
- **c.** local farms and ranches.
- **d.** local geographic features.

_______ **9.** Mountain and valley breezes are caused by
- **a.** differences in temperature and elevation.
- **b.** similarities in temperature and elevation.
- **c.** the same temperature at all elevations.
- **d.** high temperatures at all elevations.

Section Quiz

Section: The Air We Breathe

Write the letter of the correct answer in the space provided.

______ **1.** When the atmosphere is contaminated by pollutants from human and natural sources, it is called
    a. primary pollution.   **c.** air pollution.
    b. secondary pollution.   **d.** killer fog.

______ **2.** Carbon monoxide, dust, and smoke from forest fires that are put directly into the air are called
    a. smog.   **c.** secondary pollutants.
    b. primary pollutants.   **d.** killer fog.

______ **3.** Which of the following is a secondary pollutant?
    a. carbon monoxide   **c.** ozone
    b. chemicals from paint   **d.** vehicle exhaust

______ **4.** What is the major source of human-caused air pollution?
    a. animal waste   **c.** vehicle exhaust
    b. industrial chemicals   **d.** dry-cleaning businesses

______ **5.** When sulfur dioxide and nitrogen oxide are released into the air, they can cause
    a. sulfurous precipitation.   **c.** nitrogen precipitation.
    b. acid precipitation.   **d.** acidic air.

______ **6.** A rapid change in a body of water's acidity is called
    a. acid precipitation.   **c.** acid shock.
    b. acid flow.   **d.** aquatic shock.

Match the correct description with the correct term. Write the letter in the space provided.

______ **7.** allows more UV radiation to reach Earth   **a.** hybrid car

    b. air pollution

______ **8.** causes coughing, headaches, and lung cancer   **c.** ozone hole

    d. Allowance Trading System

______ **9.** limits the amount of pollution companies can release

______ **10.** uses both gasoline and electric power

Chapter Test A

The Atmosphere

MULTIPLE CHOICE

Write the letter of the correct answer in the space provided.

______ **1.** What is the atmosphere that surrounds Earth made of?
- **a.** water vapor
- **b.** carbon dioxide
- **c.** a mixture of gases
- **d.** oxygen

______ **2.** What is most of the air we breathe made of?
- **a.** oxygen
- **c.** ozone
- **b.** carbon dioxide
- **d.** nitrogen

______ **3.** What pulls gas molecules in the air toward Earth?
- **a.** air pressure
- **b.** gravity
- **c.** water
- **d.** solar energy

______ **4.** In what part of Earth's atmosphere is the thermosphere located?
- **a.** inside the troposphere
- **b.** above the mesosphere
- **c.** below the stratosphere
- **d.** below the mesosphere

______ **5.** Where does most of the human-caused air pollution come from?
- **a.** pets and other animals
- **b.** smoke from forest fires
- **c.** car exhaust
- **d.** carbon dioxide

______ **6.** What is rain, sleet, or snow that contains acids from pollution called?
- **a.** acid shock
- **b.** acid precipitation
- **c.** scattered precipitation
- **d.** acid wash

| Chapter Test A *continued*

_______ **7.** What is the main problem caused by the ozone hole?
   a. Too much radiation escapes Earth.
   b. Too much radiation reaches Earth.
   c. Too little radiation reaches Earth.
   d. Too little radiation escapes Earth.

_______ **8.** What is smog made of?
   a. ozone and vehicle exhaust
   b. ozone and killer fog
   c. smoke and vehicle exhaust
   d. smoke and ozone

MATCHING

Match the correct description with the correct term. Write the letter in the space provided.

_______ **9.** movement of air caused by differences in air pressure

_______ **10.** circular air patterns

_______ **11.** bands of high and low pressure about every 30° latitude

_______ **12.** path of the wind that curves because Earth is rotating

a. Coriolis effect

b. pressure belts

c. global winds

d. wind

Match the correct description with the correct term. Write the letter in the space provided.

_______ **13.** These flow toward the poles from west to east.

_______ **14.** This cool land air flows toward the ocean.

_______ **15.** These winds blow from the poles to 60° latitude in both hemispheres.

_______ **16.** These can blow from any direction.

a. local winds

b. westerlies

c. land breezes

d. polar easterlies

FILL-IN-THE-BLANK

Use the terms from the following list to complete the sentences below.

    radiation   conduction
    global warming   convection current

17. Energy that moves as electromagnetic waves is called

_______________________.

18. Heat moving through a material is called _______________________.

19. When the global temperature rises bit by bit, it is called

_______________________.

20. Warm air rising and cool air sinking are called a(n)

_______________________.

Name _____________________________ Class _____________ Date __________

Chapter Test B

The Atmosphere
MULTIPLE CHOICE
Write the letter of the correct answer in the space provided.

______ **1.** What is the atmosphere?
   a. oxygen
   b. carbon dioxide
   c. a mixture of gases
   d. water vapor

______ **2.** The air we breathe is mostly
   a. oxygen.
   b. carbon dioxide.
   c. ozone.
   d. nitrogen.

______ **3.** About how much of Earth's atmosphere is oxygen?
   a. 21%
   b. 78%
   c. 35%
   d. 50%

______ **4.** Why is air pressure greatest at Earth's surface?
   a. because of the pressure of oxygen
   b. because gravity pulls gas molecules toward the surface
   c. because of the weight of ice crystals
   d. because of pollution

______ **5.** Air temperature changes as altitude increases because of
   a. gases that absorb solar energy.
   b. gravity's pull on oxygen.
   c. air pollution.
   d. air pressure.

______ **6.** The protective ozone layer is found in the
   a. thermosphere.
   b. mesosphere.
   c. troposphere.
   d. stratosphere.

Chapter Test B *continued*

______ **7.** Most solar energy that reaches Earth's atmosphere is
   a. absorbed by Earth's surface.
   b. reflected by Earth's surface.
   c. scattered by clouds.
   d. absorbed by clouds, ozone, and gases.

______ **8.** Radiation is the transfer of energy
   a. as electromagnetic waves.
   b. by circulation of gases.
   c. from atmospheric gases.
   d. as heat through a material.

______ **9.** Conduction is the transfer of energy
   a. by the circulation of gases or liquids.
   b. as electromagnetic waves.
   c. as heat through a material.
   d. to the atmosphere.

______ **10.** Convection is the transfer of energy
   a. by the circulation of gases or liquids.
   b. as electromagnetic waves.
   c. as heat through a material.
   d. to the atmosphere.

______ **11.** Global warming may be caused by
   a. a decrease in greenhouse gases.
   b. an increase in greenhouse gases.
   c. the escape of thermal energy.
   d. the escape of radiation.

MATCHING

Match the correct description with the correct term. Write the letter in the space provided.

_______ **12.** movement of air caused by differences in air pressure

_______ **13.** winds that blow from 30° latitude in both hemispheres toward the equator

_______ **14.** winds that blow from 30° to 60° latitude in both hemispheres

_______ **15.** winds that blow from the poles to 60° latitude in both hemispheres

_______ **16.** winds that result from pressure differences due to properties of Earth's surface

_______ **17.** an example of a primary pollutant

_______ **18.** the largest source of human-caused air pollution in the United States

_______ **19.** what allows dangerous UV radiation to reach Earth

_______ **20.** a vehicle that reduces pollution by running on both electricity and gasoline

a. westerlies

b. wind

c. trade winds

d. polar easterlies

e. sea salt

f. local winds

g. hybrid car

h. vehicle exhaust

i. ozone hole

Chapter Test B *continued*

MATCHING

Use the figure below to answer questions 21 through 25. Write the letter of the correct answer in the space provided.

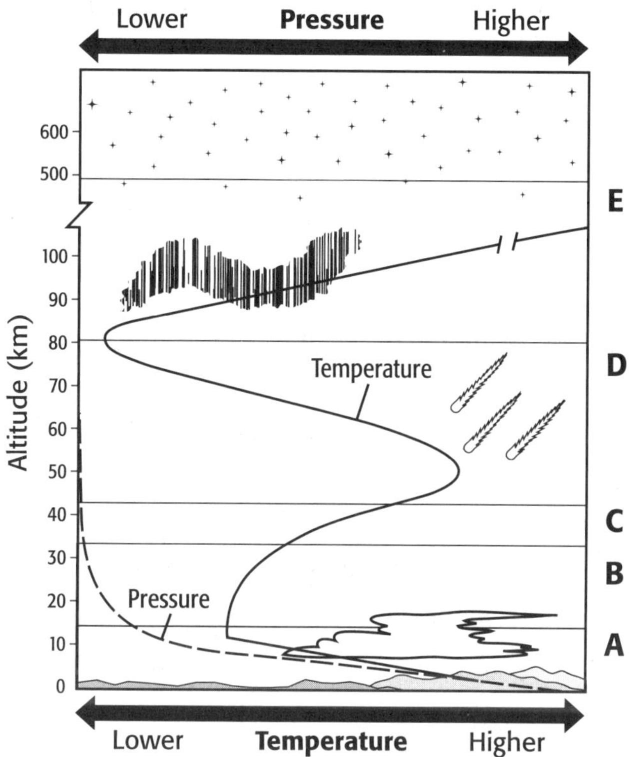

_______ **21.** the coldest layer of the atmosphere

_______ **22.** part of the atmosphere that contains the ozone layer

_______ **23.** layer of the atmosphere where temperatures can reach 1,000°C

_______ **24.** layer of the atmosphere in which weather occurs

_______ **25.** layer of the atmosphere that absorbs harmful ultraviolet radiation

a. troposphere

b. stratosphere

c. ozone layer

d. mesosphere

e. thermosphere

Chapter Test C

The Atmosphere

USING KEY TERMS

Use the terms from the following list to complete the sentences below. Each term may be used only once. Some terms may not be used.

westerlies	secondary	trade winds
greenhouse effect	Coriolis effect	conduction
polar easterlies	primary	

1. When you pick up a hot cup, heat is transferred from the cup to your hand

by ____________________.

2. Ozone and smog are examples of ____________________

pollutants.

3. Winds that flow toward the poles in the opposite direction of the trade

winds are called ____________________.

4. The ____________________ is caused by gases in the atmosphere that

absorb radiation and transfer heat.

5. Winds in the Northern Hemisphere traveling north curve to the east and

winds traveling south curve to the west due to the ____________________.

UNDERSTANDING KEY IDEAS

Write the letter of the correct answer in the space provided.

______ **6.** Wind occurs because air tends to move from regions of higher to lower
 a. latitude.
 b. pressure.
 c. nitrogen levels.
 d. humidity.

______ **7.** The protective ozone layer is found in the
 a. thermosphere.
 b. mesosphere.
 c. troposphere.
 d. stratosphere.

______ **8.** Air pressure decreases as what increases?
 a. altitude
 b. radiation
 c. water vapor
 d. pollution

9. Why do nighttime land breezes occur in areas close to the ocean?

10. Why do mountain breezes occur?

11. How can gases in Earth's atmosphere be compared to the glass covering a greenhouse?

CRITICAL THINKING

12. Suppose the stratosphere became covered in a thick blanket of volcanic dust. How would this affect the temperature of the air in the troposphere?

13. If all air pollution controls were completely removed, how do you think this would affect the environment?

14. Suppose in the future, the ozone layer disappeared completely. What impact would that have on Earth's environment?

____

____

____

____

____

INTERPRETING GRAPHICS

The graph shows atmospheric carbon dioxide levels at a site in Hawaii from 1958 to 1988. Examine the graph, and answer the question that follows.

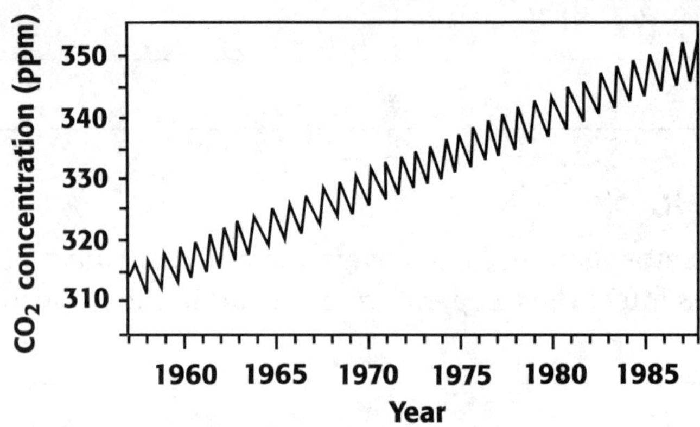

15. Why might scientists be concerned about the trend shown in the graph above?

____

____

____

____

____

____

| Chapter Test C *continued*

CONCEPT MAPPING

16. Use the following terms to complete the concept map below:

density	heat	temperature
thermosphere	nitrogen and oxygen	particles in motion
solar energy		

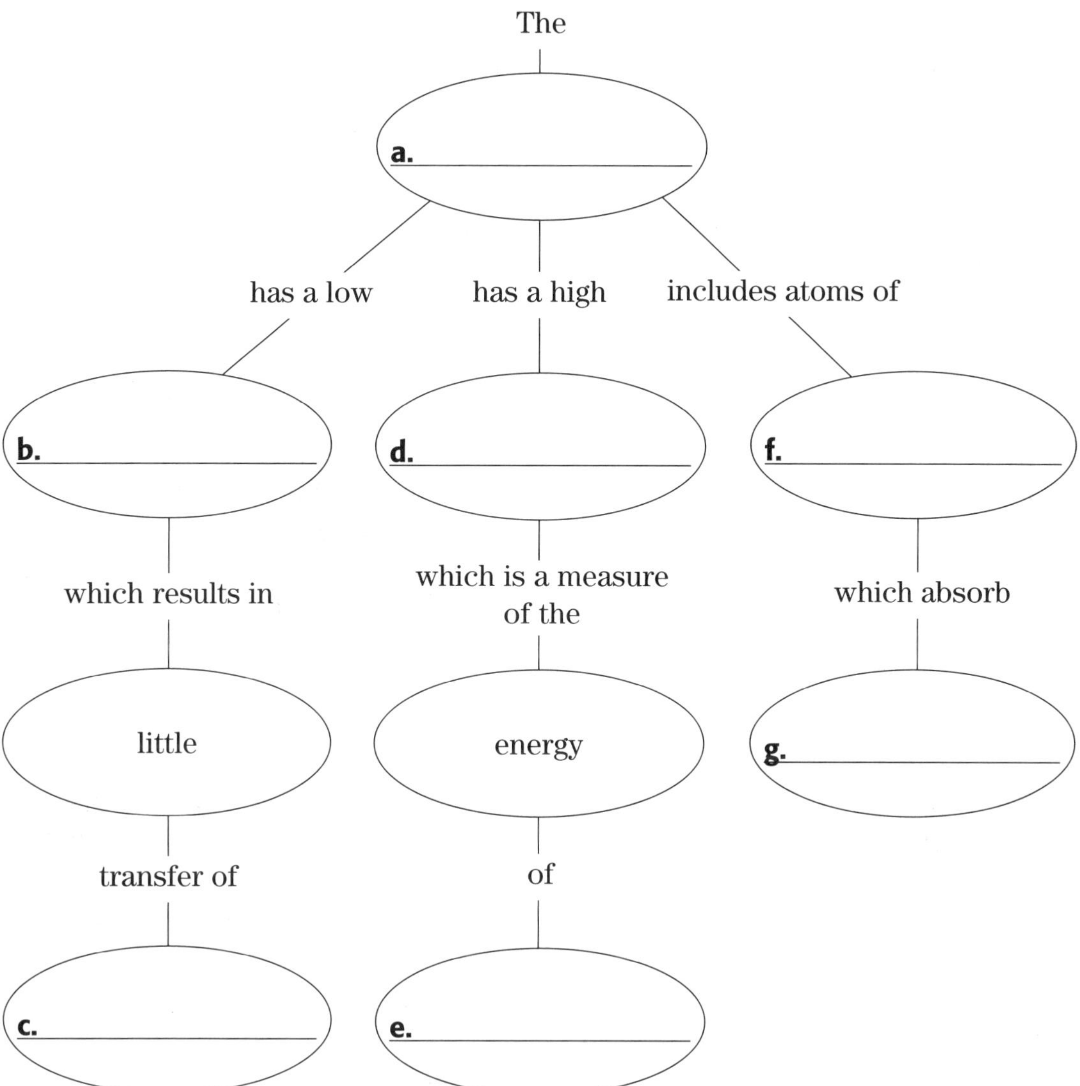

Performance-Based Assessment

Teacher Notes

PURPOSE

Students will use terraria and heat lamps to observe the effects of greenhouse gases on the heating and cooling of the atmosphere.

Bert Sherwood
Soccero Middle School
El Paso, Texas

TIME REQUIRED

One 45-minute class period. Students will need 25 minutes to perform the experiment and 20 minutes to answer the analysis questions.

RATING

Easy ←—— 1 2 3 4 ——→ Hard

Teacher Prep–2
Student Set-Up–1
Concept Level–2
Clean Up–1

ADVANCE PREPARATION

Assemble terraria by cutting the tops off plastic water jugs or similar containers. Equip each activity station with two terraria, each with a layer of sand in the bottom. Attach or suspend a thermometer in the same location inside each terrarium. Place the terraria under the heat lamps, but do not turn the lamps on. Make sure the lamps are positioned at the same height, at least 30 cm above the terraria.

SAFETY CAUTION

Remind students to review all safety cautions and icons before beginning this activity. Instruct students not to touch the bulbs of the heat lamps. Use alcohol thermometers rather than mercury thermometers. Instruct students not to touch broken thermometers. Have a disposal container available in case thermometers break.

TEACHING STRATEGIES

This activity works best when performed in pairs.

Performance-Based Assessment *continued*

Evaluation Strategies

Use the following rubric to help evaluate student performance.

Rubric for Assessment

Possible points	Appropriate use of materials and equipment (30 points possible)
30–20	Successfully completes activity; safe and careful handling of materials and equipment; attention to detail; superior lab skills
19–10	Task is generally complete; successful use of materials and equipment; sound knowledge of lab techniques; somewhat unfocused performance; mild neglect of safety measures
9–1	Attempts to complete activity; inadequate results; unsafe lab technique; apparent lack of skill
	Quality and clarity of observations (40 points possible)
40–30	Superior observations stated clearly and accurately; high level of detail; correct use of units of measurement
29–20	Accurate observations; moderate level of detail; correct use of units of measurement
19–10	Observations are complete but expressed in unclear manner, may include minor inaccuracies; attempts to use units of measurement include errors or inconsistencies
9–1	Erroneous, incomplete, or unclear observations; lack of accuracy, details, units of measurement
	Explanation of observations (30 points possible)
30–20	Clear, detailed explanation that shows superior understanding of greenhouse effect and atmospheric heat transfer
19–10	Adequate understanding of greenhouse effect and atmospheric heat transfer; minor difficulty in expression
9–1	Poor understanding of greenhouse effect and atmospheric heat transfer; explanation unclear or not relevant; substantial factual errors

  SKILLS PRACTICE

Performance-Based Assessment

OBJECTIVE

You have learned about the atmosphere, including factors that affect heating, cooling, and the temperature of air. In this activity, you will demonstrate the heating of the atmosphere. You will also see how the greenhouse effect can affect atmospheric temperatures.

KNOW THE SCORE!

As you work through the activity, keep in mind that you will be earning a grade for the following:

- how well you work with materials and equipment (30%)

- how clearly you make your observations (40%)

- how well you use your knowledge of the atmosphere and the greenhouse effect to explain your observations (30%)

MATERIALS AND EQUIPMENT

- beaker, 200 mL
- cover for one terrarium, clear plastic
- heat lamps (2)
- sand

- terraria (2)
- thermometers (2)
- watch or timer
- water

SAFETY INFORMATION 

- Do not touch the bulbs of the heat lamps.
- Use alcohol rather than mercury thermometers.
- Do not touch broken thermometers.

PROCEDURE

1. Place sand in each terrarium, and then pour 200 mL water onto the sand. Place a thermometer in each terrarium, making sure that you can read the temperature. Then, cover one terrarium with plastic.

2. Turn on the two heat lamps at the same time.

3. Read the temperature for each terrarium every other minute for 10 minutes, and record the data in the table on the next page.

4. Turn off the heat lamps after 10 minutes. Continue to record the temperature every other minute for another 10 minutes.

Performance-Based Assessment *continued*

ANALYZE THE RESULTS

5. In which terrarium did the temperature increase faster, the one with the cover or the one without the cover?

____

____

6. How does the cover affect the cooling of the terrarium?

____

____

____

Data Table		
Time (min)	**Covered (°C)**	**Uncovered (°C)**

Performance-Based Assessment *continued*

7. Do you think the temperature changes would have been different if there had been no water in the terrarium? Explain.

____

____

____

____

8. Which terrarium do you think is a better model for Earth's atmosphere? Why?

____

____

____

____

BIG IDEA QUESTION

9. How does the ability of Earth's atmosphere to trap heat, which the experiment demonstrated, help explain how the atmosphere makes Earth livable?

____

____

____

____

____

____

____

Assessment)

Standards Assessment

Teacher Notes and Answer Key

To provide practice under more realistic testing conditions, give students 20 min to answer all of the questions in this assessment.

QUESTION NUMBER	CORRECT ANSWER	STANDARD
1	*A*	6.4.a (supporting)
2	*B*	6.3.a (supporting)
3	*D*	6.4.d (supporting)
4	*C*	6.4.b (supporting)
5	*C*	6.3.d (mastering)
6	*C*	6.6.a (mastering)
7	*C*	6.4.e (mastering)
8	*C*	6.3.c (supporting)
9	*C*	6.6.a (supporting)
10	*D*	6.4.e (supporting)
11	*B*	6.6.a (supporting)
12	*D*	6.6.a (supporting)
13	*D*	5.3.c (mastering)
14	*A*	5.4.c (mastering)

TEST DOCTOR

The following Standards Assessment questions have been diagnosed by the Test Doctor. Find out what might be causing your students' "ailing" answers. Each Test Doctor is followed by a diagnostic teaching tip to help you address students' learning needs.

Question 1 *asks students to demonstrate an understanding of the meaning of the word* energy.

A **Correct.** Energy is the power that makes things happen, whether in your home, your body, or the world around you.

B **Incorrect.** Many things happen within cycles, yet cycle is not the power that makes things happen.

C **Incorrect.** Data is information. It is not the power to make things happen.

D **Incorrect.** A factor is a general term that can be variable. It is not the power to make things happen.

Diagnostic Teaching Tip: Students who have difficulty answering this question correctly might benefit from a review of the meanings of all the words presented here. Once students are comfortable with the definitions, have groups of students create short skits that use all of the words in conversation.

Standards Assessment *continued*

Question 2 *asks students to demonstrate an understanding of the meaning of the word* involve.

A Incorrect. A circle is a fairly inclusive shape, but a circle does not mean "to have as a part of."

B Correct. To involve a thing is to have it as a part of something.

C Incorrect. You may have to replace a part of something, but replace does not mean "to have as a part of."

D Incorrect. *Relate* means "to have a connection with"; it does not mean "to have as a part of."

Diagnostic Teaching Tip: Students who have difficulty answering this question correctly might benefit from using various forms of *involve* in context. Have students brainstorm situations in which things are involved with other things.

Question 3 *asks students to identify the meaning of the word* distribute.

A Incorrect. *Require* and *distribute* are not synonymous.

B Incorrect. Convection currents do not wear away heat.

C Incorrect. When something is distributed, it may be changed, but the two terms are not synonymous.

D Correct. Heat is distributed, or spread out, by convection currents in the atmosphere.

Diagnostic Teaching Tip: Students who struggle with this question might benefit from substituting each verb presented here with the verb *distribute*. Ask students which statement remains true. If students are still unable to find the correct answer, have them review the process of heat transfer by convection.

Question 4 *asks students to choose a phrase that means* visible.

A Incorrect. Visible objects do not necessarily make sound.

B Incorrect. Some things that are not visible can still be measured.

C Correct. Visible objects can be seen.

D Incorrect. A visible object can be selected, but smells and sounds can be used to make selections of things that are not visible.

Diagnostic Teaching Tip: Students who answer incorrectly may not have read all the choices carefully, or may not have read the question carefully. Something that is visible might be measured, or heard, or selected, but these words are not close in meaning to *visible*.

Question 5 *asks students to demonstrate understanding of how the sun's energy reaches Earth.*

A Incorrect. Most of the sun's radiation does not reach the surface.

B Incorrect. Most of the visible light reaches Earth's surface.

C Correct. Most radiation is absorbed, but most visible light is not.

D Incorrect. Infrared radiation does reach the surface undisturbed.

Diagnostic Teaching Tip: Students who have difficulty answering this question correctly might benefit from reviewing the properties of the electromagnetic spectrum. Have groups of students create skits demonstrating what parts of the atmosphere absorb ultraviolet rays, X rays, gamma rays, and infrared waves.

Question 6 *asks students to demonstrate understanding of the consequences of fossil fuel consumption.*

A Incorrect. Acid rain does not cause global warming.
B Incorrect. Global warming is not caused by a breakdown in the atmosphere.
C Correct. Fossil fuels release carbon dioxide, which traps heat in the atmosphere, causing global warming.
D Incorrect. Blocking sunlight would not contribute to global warming.

Diagnostic Teaching Tip: Students who struggle to answer this question correctly might benefit from reviewing the effects of greenhouse gases. For example, discuss how a greenhouse works to help plants grow. Ask students to hypothesize how greenhouse gases, such a carbon dioxide, work within Earth's atmosphere. Have students draw comparative diagrams of each process, showing the effects of each.

Question 7 *asks students to describe how changes in air pressure change weather patterns.*

A Incorrect. The speed of wind does not increase as pressure equalizes.
B Incorrect. Wind does not blow toward an area of high pressure.
C Correct. As the air pressure in Area B increases, the wind blowing toward Area B begins to move more slowly.
D Incorrect. Wind does not blow toward Area A in this scenario.

Diagnostic Teaching Tip: Students who have difficulty answering this question correctly might benefit from reviewing the behavior of air currents in high and low pressure areas. Using an open window or a fan, create an area of high pressure on one side of the door to the classroom. Have students experiment feeling the air pressure change in the hallway when the door is opened. Does the force of the wind change when the door is opened only a little, versus when it is opened wide? Does the force of the wind change when the door remains open for an extended period of time?

Question 8 *asks students to demonstrate understanding that temperature differences cause air movements.*

A Incorrect. Air movements from the ocean are not a result of Earth's rotation.
B Incorrect. Ocean storms are not the primary reason that air moves toward land.
C Correct. Cooler air from over the ocean moves toward land to replace air that has heated and risen.
D Incorrect. If the air over land heated more slowly than the air over the ocean, it would not rise and would therefore not be replaced by air over the ocean.

Standards Assessment *continued*

Diagnostic Teaching Tip: Students who have difficulty with this question might benefit from using memory tools to remember how the direction of air movement over land and the ocean changes from night to day. Encourage students to create a short rhyme or mnemonic device to remind themselves that wind moves toward land during the day and toward the ocean at night.

Question 9 *asks students to demonstrate understanding of the causes and effects of air pollution.*

A Incorrect. Cars produce 10 to 20 percent of air pollution, not a majority.

B Incorrect. Air pollution can occur indoors as well as outdoors.

C Correct. Air pollution caused by the consumption of fossil fuels can result in acid precipitation.

D Incorrect. The use of hybrid cars helps to reduce air pollution levels.

Diagnostic Teaching Tip: Students who have difficulty answering this question correctly might benefit from identifying sources of air pollution and noting their effects locally. Have students observe and research to identify sources of air pollution in your area. Invite students to share the information they have gathered. Create a pie chart showing their data.

Question 10 *asks students to demonstrate understanding of how pressure and temperature fluctuate in the atmosphere.*

A Incorrect. Temperature does not decrease consistently with higher altitudes.

B Incorrect. Increased altitudes do not lead to increased temperatures and pressure.

C Incorrect. Temperature does not increase consistently as altitude increases.

D Correct. As altitude increases, pressure decreases and temperature fluctuates.

Diagnostic Teaching Tip: Students who have difficulty with this question might benefit from creating a chart of pressure and temperature change at different altitudes of the atmosphere. Have students create a diagram of the layers of the atmosphere and label the layers with their names, the level of atmospheric pressure, and the range of temperature.

Question 11 *asks students to demonstrate understanding of the consequences of using pollutants.*

A Incorrect. Pollen is a natural substance. Except under conditions of unusual concentration, pollen is not classified as a pollutant.

B Correct. Because smog is formed from a reaction between a primary pollutant—vehicle exhaust—and sunlight, it is classified as a secondary pollutant.

C Incorrect. Paint chemicals are, themselves, primary pollutants.

D Incorrect. Vehicle exhaust is a primary pollutant.

Diagnostic Teaching Tip: Students who have difficulty answering this question correctly might benefit from practice identifying primary and secondary pollutants. Create a graphic organizer to classify the information, and then have students brainstorm examples for each category.

| Standards Assessment *continued*

Question 12 *asks students to identify the consequences of a damaged ozone layer.*

A Incorrect. Ozone layer depletion does not produce acid rain.
B Incorrect. The thinning ozone layer does not intensify the effects of pollutants.
C Incorrect. The greenhouse effect is created by carbon dioxide emissions, not the thinning ozone layer.
D Correct. The thinning of the ozone layer reduces our protection from ultraviolet radiation from the sun, which can cause skin cancer.

Diagnostic Teaching Tip: Students who struggle with this question might benefit from creating a chart describing various environmental degradations. Have groups of students list the causes and effects of the greenhouse effect, the ozone hole, and acid rain. Then, have students choose one problem to target in a poster encouraging ways to alleviate the problem.

Question 13 *asks students to demonstrate understanding of the process of condensation.*

A Incorrect. Water vapor is in a gaseous state and therefore cannot evaporate.
B Incorrect. Cooled water vapor does not necessarily move to a lower-pressure area.
C Incorrect. In this scenario, water vapor does not move to a higher-pressure area.
D Correct. Water vapor that cools is likely to condense and fall as precipitation.

Diagnostic Teaching Tip: Students who have difficulty answering this question correctly might benefit from reviewing the steps in the water cycle. Have students create a circular diagram showing evaporation, condensation, precipitation, and runoff. Then, divide students into groups, and instruct each student to describe what happens in one step of the cycle.

Question 14 *asks students to demonstrate understanding of the origins of severe weather patterns.*

A Correct. When high-pressure areas surround a low-pressure area, the low-pressure tends to rise and be replaced by high-pressure air from all around. Earth's rotation deflects these winds, which has a net effect of a circular wind. This effect causes a cyclone.
B Incorrect. Severe weather patterns do not result when low-pressure areas surround a high-pressure area.
C Incorrect. A low-pressure area and a high-pressure area, side by side, do not create severe weather.
D Incorrect. This air pressure pattern does not create severe weather.

Diagnostic Teaching Tip: Students who have difficulty answering this question correctly might benefit from reviewing the steps involved in the formation of severe weather. Challenge students to create a flowchart that shows air from the low-pressure zone moving and being replaced by the high-pressure air, rotating due to Earth's rotation, and creating a cyclone that leads to severe weather.

Assessment

Standards Assessment

REVIEWING ACADEMIC VOCABULARY

______ **1.** Which of the following words means "the power that makes things happen"?
  A energy
  B cycle
  C data
  D factor

______ **2.** Which of the following words means "to have as a part of"?
  A circle
  B involve
  C replace
  D relate

______ **3.** In the sentence "Convection currents distribute heat in the atmosphere," what does the word *distribute* mean?
  A to require
  B to wear away
  C to change
  D to spread out

______ **4.** Which of the following phrases is the closest in meaning to the word *visible*?
  A can be heard
  B can be measured
  C can be seen
  D can be selected

REVIEWING CONCEPTS

______ **5.** How does Earth's atmosphere affect radiation that travels from the sun?
  A Almost all the radiation reaches the surface of Earth.
  B Most of the visible light is absorbed by the atmosphere.
  C Most visible light reaches the surface of Earth.
  D Only infrared radiation reaches the surface undisturbed.

______ **6.** How might burning fossil fuels contribute to global warming?
  A by polluting water that then becomes acid precipitation in the water cycle
  B by breaking down the atmosphere and allowing more radiation in
  C by producing carbon dioxide that traps heat in the atmosphere
  D by producing pollution that blocks sunlight from reaching Earth

Standards Assessment *continued*

______ **7.** Area A, which has high air pressure, is located beside Area B, which has low air pressure. The wind is blowing toward Area B. Which of the following is most likely to occur as the air pressure in Area B increases?

    A The wind blowing toward Area B becomes faster.
    B The wind blowing toward Area A becomes faster.
    C The wind blowing toward Area B becomes slower.
    D The wind blowing toward Area A becomes slower.

______ **8.** During the day, air over the ocean blows toward the land. Which of the following explains this phenomenon as shown in the figure above?

    A Earth's rotation causes air to blow toward land.
    B The energy of ocean storms pushes air toward land.
    C Air over land heats more quickly, so the air rises and is replaced by air from the ocean.
    D Air over land heats more slowly, so the warmer air over the ocean moves in to replace it.

______ **9.** Which of the following statements about air pollution is true?

    A Cars and trucks produce the majority of air pollution.
    B Air pollution does not occur indoors; it occurs only outdoors.
    C Air pollution caused by fossil-fuel use can lead to acid precipitation.
    D The use of hybrid cars has increased levels of air pollution.

______ **10.** Which of the following statements describes how pressure and temperature change as altitude increases?

    A Pressure and temperature decrease.
    B Pressure and temperature increase.
    C Pressure decreases and temperature increases.
    D Pressure decreases and temperature fluctuates.

REVIEWING PRIOR LEARNING

______ **11.** Which of the following is an example of a secondary pollutant?

    A pollen
    B smog
    C paint chemicals
    D vehicle exhaust

| Standards Assessment *continued*

_______ **12.** What effect does the thinning of the ozone layer have on
the environment?
  A It creates acid precipitation that harms the environment.
  B It intensifies the effects of primary and secondary pollutants.
  C It creates a greenhouse effect that leads to global warming.
  D It increases ultraviolet radiation that can cause skin cancer.

_______ **13.** Which of the following is most likely to occur when water vapor in a
cloud cools?
  A It evaporates and rises.
  B It moves away to a lower-pressure area.
  C It moves away to a higher-pressure area.
  D It condenses and falls as precipitation.

A	**B**	**C**	**D**
H H	L L	L H	H L
H LL H	L H H L	L H	L H
HH	L L	L H	

_______ **14.** In the air-pressure patterns above, H represents high pressure, and L
represents low pressure. Which of the air-pressure patterns is most
likely to lead to severe weather, such as a hurricane?
  A pattern A
  B pattern B
  C pattern C
  D pattern D

Sunlight and Temperature Change

Teacher Notes

In this activity, students measure the effects of direct sunlight on temperature (covers standards 6.4.b and 6.7.b).

MATERIALS

For each group

- construction paper or cardstock
- mirror, plastic
- modeling clay
- thermometers, outdoor (2)

SAFETY CAUTION

Remind students to review all safety cautions and icons before beginning this activity. Caution students not to look directly into the sun or its reflection.

Sunlight and Temperature Change

In this activity, you will observe a relationship between direct sunlight and temperature change.

PROCEDURE

1. Place **two outdoor thermometers** side-by-side in **clay** bases outside.

2. Create **sunshades** to block direct sunlight from falling on each thermometer.
   - Create a sunshade by folding and placing the construction paper or cardstock behind the thermometers.
   - Use the clay base to help hold the paper in place.

3. Record the temperature displayed on each shaded thermometer.
   - Thermometer 1 _______________________
   - Thermometer 2 _______________________

4. Position a **plastic mirror** so that it reflects sunlight directly onto one of the thermometers.
   - **Caution:** Do not look directly into the sun or its reflection.

5. After 15 minutes, record the temperature readings on both thermometers.
   - Thermometer 1 _______________________
   - Thermometer 2 _______________________

ANALYSIS

6. Was there any change in temperatures between the two thermometers?

   ____

   - If yes, what was the difference? Which one was warmer?

   ____

7. What is the source of the energy that produced this change?

   ____

   - How did the energy reach the thermometer that had the mirror reflecting on it?

   ____

   - How did the energy reach the thermometer that did not have the mirror reflecting on it?

   ____

8. How might additional mirrors focusing light on the thermometer change your results?

   ____

   ____

Sunlight and Temperature Change

In this activity, you will observe a relationship between direct sunlight and temperature change.

PROCEDURE

1. Position **two outdoor thermometers** side by side in **clay** bases, and place them outside.

2. Create **sunshades** to block direct sunlight from falling on each thermometer.

3. Record the temperature displayed on each shaded thermometer.

____

4. Position a **plastic mirror** so that it reflects sunlight directly onto one of the thermometers. **Caution:** Do not look directly into the sun or its reflection.

5. After 15 min, read the thermometers and record their temperatures.

____

ANALYSIS

6. Was there any change in observed temperatures? Describe your observations.

____

____

7. What is the source of the energy that produced this change? How did the energy reach the thermometers?

____

____

8. How might additional mirrors focusing light on the thermometer affect your results?

____

____

DATASHEET C

Sunlight and Temperature Change

In this activity, you will observe a relationship between direct sunlight and temperature change.

PROCEDURE

1. Position **two outdoor thermometers** side-by-side in **clay** bases and place them outside.

2. Create **sunshades** to block direct sunlight from falling on each thermometer.

3. Record the temperature displayed on each shaded thermometer.

4. Position a **plastic mirror** so that it reflects sunlight directly onto one of the thermometers. **Caution:** Do not look directly into the sun or its reflection.

5. After 15 min, record the temperature readings on both thermometers.

ANALYSIS

6. Was there any change in observed temperatures? Describe your observations.

____

____

____

7. What is the source of the energy that produced this change? How did the energy reach each thermometer?

____

____

____

8. How might additional mirrors affect your results? Is there a way that additional mirrors might decrease the amount of energy reaching the thermometer? Explain your answer.

____

____

____

Modeling Air Pressure

Teacher Notes

This activity has students use a cup of water and an index card to demonstrate that air exerts pressure in all directions (covers standard 6.4.e).

MATERIALS

For each group

- cup, plastic
- index cards
- sink or bucket
- water

SAFETY CAUTION

Remind students to review all safety cautions and icons before beginning this activity. Students should keep the glass of water over a sink when flipping the glass.

Modeling Air Pressure

SAFETY INFORMATION

PROCEDURE

1. Fill a **plastic cup** to the brim with **water.**
2. Firmly hold an **index card** over the mouth of the cup.
3. Quickly invert the glass over a **sink.**
4. Let go of the index card.
   - Watch what happens.
   - Does the water spill out of the cup? Why or why not?

   ____

   ____

   ____

5. The index card is being held up by _______________________.
   - The force of air pressure holding the index card up must be

   _______________________ than the force of gravity pulling down
   on the water.

Modeling Air Pressure

SAFETY INFORMATION

PROCEDURE

1. Fill a **plastic cup** to the brim with **water.**

2. Firmly hold an **index card** over the mouth of the cup.

3. Quickly invert the glass over a **sink.**

4. Release the paper square, and observe what happens.

5. How do the effects of air pressure explain your observations?

____

____

____

____

____

<table><tr><td>Quick Lab</td><td align="right">**DATASHEET C**</td></tr></table>

Modeling Air Pressure

SAFETY INFORMATION

PROCEDURE

1. Fill a **plastic cup** to the brim with **water.**

2. Firmly hold an **index card** over the mouth of the cup.

3. Quickly invert the glass over a **sink.**

4. Release the index card, and observe what happens. Write your observations below.

____

____

____

____

5. What force caused what you observed? How do the effects of this force explain your observations?

____

____

____

____

  DATASHEET

Modeling Air Movement by Convection

Teacher Notes

This activity has students use ice and a lamp to model the formation of convection currents (covers standards 6.4.d, 6.7.e, and 6.7.g).

MATERIALS

For each group

- aquarium, glass, 15 gal
- ice cubes (24)
- incense stick
- lamp, adjustable neck
- masking tape
- plastic cling wrap
- thermometers, outdoor (2)

SAFETY CAUTION

Remind students to review all safety cautions and icons before beginning this lab.

Quick Lab **DATASHEET A**

Modeling Air Movement by Convection

In this lab, you will model how the transfer of energy causes convection currents to form.

SAFETY INFORMATION

TRY IT!

1. Use **masking tape** to tape **two outdoor thermometers** to the inside of a **15 gal glass aquarium,** one at each end.
  • Make sure the thermometers can be read from outside the aquarium.

2. Stack **24 ice cubes** in one end of the aquarium.

3. Place an **adjustable neck lamp** outside the other end of the aquarium.
  • Point the lamp at the bottom of that end of the aquarium.

4. Tape a piece of **plastic cling wrap** over the top of the aquarium
  • Wait 5 minutes.

5. Write down the temperature on each thermometer.
  • Temperature near the ice:

  • Temperature near the lamp:

6. Light the end of an **incense stick.** Then, quickly blow out the flame.

7. Lift the plastic wrap, and hold the incense stick in the center of the aquarium.
  • Note the direction in which the smoke moves in relation to the lamp and the ice cubes.

THINK ABOUT IT!

8. Why does the smoke move in the direction it does? (Hint: Think about how warm air moves compared to cold air.)

____

____

____

____

____

____

Modeling Air Movement by Convection *continued*

9. How does this activity model convection in the atmosphere? (Remember that convection currents may be vertical, circular, or cyclical.)

____

____

____

____

____

____

Modeling Air Movement by Convection

In this lab, you will model how the transfer of energy causes convection currents to form.

SAFETY INFORMATION

TRY IT!

1. Use **masking tape** to attach **two outdoor thermometers** to the inside of a **15 gal glass aquarium,** one at each end. Make sure the thermometers can be read from outside the aquarium.

2. Stack **24 ice cubes** in one end of the aquarium.

3. Place an **adjustable neck lamp** outside the other end of the aquarium. Point the lamp at the bottom of that end of the aquarium.

4. **Tape** a piece of **plastic cling wrap** over the top of the aquarium, and wait 5 min.

5. Write down the temperature on each thermometer.

6. Light the end of an **incense stick.** Then, quickly blow out the flame.

7. Lift the plastic wrap, and hold the incense stick in the center of the aquarium. Observe the movement of the smoke.

THINK ABOUT IT!

8. How does the temperature at each end of the aquarium affect the movement of the smoke?

____

____

9. How does this activity model convection in the atmosphere?

____

____

____

Modeling Air Movement by Convection

In this lab, you will model how the transfer of energy causes convection currents to form.

SAFETY INFORMATION

TRY IT!

1. Use **masking tape** to attach **two outdoor thermometers** to the inside of a **15 gal glass aquarium,** one at each end. Make sure the thermometers can be read from outside the aquarium.

2. Stack **24 ice cubes** in one end of the aquarium.

3. Place an **adjustable neck lamp** outside the other end of the aquarium. Point the lamp at the bottom of that end of the aquarium.

4. **Tape** a piece of **plastic cling wrap** over the top of the aquarium, and wait 5 min.

5. Write down the temperature on each thermometer.

____

____

6. Light the end of an **incense stick.** Then, quickly blow out the flame.

7. Lift the plastic wrap, and hold the incense stick in the center of the aquarium. Observe the movement of the smoke.

THINK ABOUT IT!

8. Why does the smoke move the way that it does?

____

____

____

____

____

____

Modeling Air Movement by Convection *continued*

9. How does this activity model convection in the atmosphere?

____

____

____

____

____

____

DATASHEET

Investigating the Coriolis Effect

Teacher Notes

This activity has students model the Coriolis effect (covers standards 6.4.e and 6.7.b).

MATERIALS

For each group

- cardboard disk, 10 cm radius
- pencil, sharpened
- scissors
- water drops

SAFETY CAUTION

Remind students to review all safety cautions and icons before beginning this activity.

Name _________________________________ Class ______________ Date _____________

Investigating the Coriolis Effect

SAFETY INFORMATION

PROCEDURE

1. Use **scissors** to cut out a **cardboard** circle that has a diameter of 20 cm.

2. Insert a **pencil** through the center of the disk.
- Place the tip of the eraser on a table so that the cardboard disk is tilted slightly.

3. Place a few drops of **water** near the center of the disk, and spin the disk on the pencil tip.
- What path do the water droplets follow as they flow off the disk?

____

____

4. If the cardboard disk is Earth, and the pencil is Earth's rotational axis, how does the movement of the water on the disk model the Coriolis effect on Earth's ocean currents and global winds?

____

____

____

____

____

____

DATASHEET B

Quick Lab

Investigating the Coriolis Effect

SAFETY INFORMATION

PROCEDURE

1. Use **scissors** to cut out a **cardboard** disk that has a diameter of 20 cm.

2. Insert a **pencil** through the center of the disk. Place the tip of the eraser on a table so that the cardboard disk is tilted slightly.

3. Place a few drops of **water** near the center of the disk, and spin the disk on the pencil tip. What happens?

____

____

____

____

4. How does the motion of the water model the Coriolis effect?

____

____

____

____

Quick Lab

DATASHEET C

Investigating the Coriolis Effect

SAFETY INFORMATION

PROCEDURE

1. Use **scissors** to cut out a **cardboard** disk that has a diameter of 20 cm.

2. Insert a **pencil** through the center of the disk. Place the tip of the eraser on a table so that the cardboard disk is tilted slightly.

3. Place a few drops of **water** near the center of the disk, and spin the disk on the pencil tip. Describe what happens.

____

____

4. How does the motion of the water model the Coriolis effect?

____

____

____

- What happens if you spin the pencil in the opposite direction from your first experiment? How is this related to what happens on Earth?

____

____

____

____

Quick Lab

DATASHEET

Collecting Air-Pollution Particles

Teacher Notes

This activity has students collect particulates to study the types of air pollution around your school (covers standards 6.6.a and 6.7.a).

MATERIALS

For each group

- hole punch
- index cards, 5 in. × 7 in. (10)
- magnifying lens
- petroleum jelly
- string

Quick Lab

Collecting Air-Pollution Particles

PROCEDURE

1. Cover **ten 5 in. × 7 in. index cards** with a thin coat of **petroleum jelly.**

2. Use **string** to hang the cards in various locations inside and outside your school.

3. One day later, use a **magnifying lens** to count the number of particles on the cards.

4. Which location had the fewest particles?

   ____

   • Which location had the most particles?

   ____

5. Why might some locations have a higher number of particles, or more air pollution, than other locations did?

   ____

   ____

   ____

   ____

   ____

Collecting Air-Pollution Particles

PROCEDURE

1. Cover **ten 5 in. × 7 in. index cards** with a thin coat of **petroleum jelly.**

2. Hang the cards in various locations inside and outside your school.

3. One day later, use a **magnifying lens** to count the number of particles on the cards.

4. Which location had the fewest particles? Which location had the most particles?

____

____

____

5. Hypothesize why some locations had a higher number of particles than other locations did.

____

____

____

____

Collecting Air-Pollution Particles

PROCEDURE

1. Cover **ten 5 in. × 7 in. index cards** with a thin coat of **petroleum jelly.**

2. Hang the cards in various locations inside and outside your school.

3. One day later, use a **magnifying lens** to count the number of particles on the cards.

4. Which location had the fewest particles? Which location had the most particles?

____

____

5. Hypothesize why some locations had a higher number of particles than other locations did.

____

____

____

• Were there any areas where the amount of pollutants surprised you? Were there more pollutants or fewer pollutants than you expected? Explain.

____

____

____

• Do you think the pollutants you captured in the petroleum jelly were primary or secondary pollutants? Explain your reasoning.

____

____

____

DATASHEET

Under Pressure!

Teacher Notes

This lab has students make a barometer and determine
how weather is affected by changes in air pressure (covers
standards 6.4.e, 6.7.a, and 6.7.e)

Carol Lindstrom
Leonard Herman
Intermediate School
San Jose, California

TIME REQUIRED

One 45-minute class period plus 15 min each day for 3 or 4 days

LAB RATINGS

Easy ←——— 1   2   3   4 ——→ Hard

Teacher Prep–2
Student Setup–4
Concept Level–2
Clean Up–2

MATERIALS

The materials listed on the student page are sufficient for a group of two to four
students.

SAFETY CAUTION

Remind students to review all safety cautions and icons before beginning this
activity.

PREPARATION NOTES

A week before the activity, have students bring in large coffee cans or jars. For
more accurate results, have students place their barometers in a shaded area.
As students work on this lab in class, have them collect newspaper clippings or
online reports of daily weather conditions.

Under Pressure!

Imagine that you are planning a picnic with your friends. The temperature this afternoon should be in the low 80s, and the skies should be sunny. These weather conditions sound quite comfortable. But you know that weather can change when air pressure changes. You can use a barometer to keep track of air-pressure changes. In this activity, you will build your own barometer and will discover what this tool can tell you.

OBJECTIVES

Predict how changes in air pressure affect weather.

Build a barometer to test your hypothesis.

MATERIALS

- balloon
- can, coffee, large, empty, 10 cm in diameter
- card, index

- scissors
- straw, drinking
- tape, masking, or rubber band

SAFETY INFORMATION 

Using Scientific Methods

ASK A QUESTION

1. What does a barometer measure?
   - How can I use a barometer to notice changes in the weather?

FORM A HYPOTHESIS

2. Complete the hypothesis below so that it answers the questions above.

   If a barometer is built correctly, it will show changes in

   _________________________, and these changes can be used to predict weather.

   If the pressure goes _________________________, the weather will be sunny and

   dry. If the pressure goes _________________________, the weather will be

   cloudy and rainy.

▎Under Pressure! *continued*

TEST THE HYPOTHESIS

3. Stretch the balloon a few times.
- Then, blow up the balloon, and let the air out.
- This step will make your barometer more sensitive to changes in atmospheric pressure.

4. Cut off the open end of the balloon.
- Stretch the balloon over the open end of the coffee can.
- Attach the balloon to the can with masking tape or a rubber band.

5. Cut one end of the straw at an angle to make a pointer.

6. Place the straw on the stretched balloon so that the pointer is directed away from the center of the balloon.
- Five centimeters of the end of the straw should extend past the edge of the can.
- Tape the straw to the balloon as shown in the illustration at right.

7. Tape the index card to the side of the can as shown in the illustration at right.
- Congratulations! You have just made a barometer!

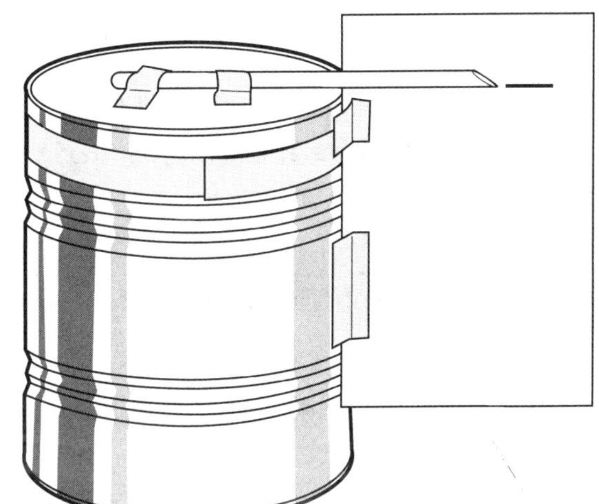

8. Now, use your barometer to collect and record information about air pressure.
- Place the barometer outside for 3 or 4 days.
- On each day, mark where the tip of the straw points on the index card.

ANALYZE THE RESULTS

9. Explaining Events Name two weather conditions that would affect how your barometer works. Explain your answer.

____

____

____

10. Recognizing Patterns Is atmospheric pressure increasing or decreasing when the straw moves up?

____

Under Pressure! *continued*

11. Recognizing Patterns Is atmospheric pressure increasing or decreasing when the straw moves down?

____

DRAW CONCLUSIONS

12. Applying Conclusions Compare your results with the barometric pressures listed in your local newspaper.
- What kind of weather occurs when high pressure is present?

____

- What kind of weather occurs when low pressure is present?

____

13. Evaluating Results If the atmospheric pressure went up, was the weather sunny and dry?

____

- If the atmospheric pressure went down, was the weather cloudy and rainy?

____

- Does the barometer you built support your hypothesis?

____

BIG IDEA QUESTION

14. Making Inferences Why is it important to be aware of barometric pressure?

____

____

____

____

- How can air pressure help scientists to predict weather?

____

____

____

____

Skills Practice Lab **DATASHEET B**

Under Pressure!

Imagine that you are planning a picnic with your friends. The temperature this afternoon should be in the low 80s, and the skies should be sunny. These weather conditions sound quite comfortable. But you know that weather can change when air pressure changes. You can use a barometer to keep track of air-pressure changes. In this activity, you will build your own barometer and will discover what this tool can tell you.

OBJECTIVES

Predict how changes in air pressure affect weather.

Build a barometer to test your hypothesis.

MATERIALS

- balloon
- can, coffee, large, empty, 10 cm in diameter
- card, index
- scissors
- straw, drinking
- tape, masking, or rubber band

SAFETY INFORMATION 

Using Scientific Methods

ASK A QUESTION

1. How can I use a barometer to detect changes in the weather?

FORM A HYPOTHESIS

2. Write a few sentences that answer the question above.

____

____

____

TEST THE HYPOTHESIS

3. Stretch the balloon a few times. Then, blow up the balloon and let the air out. This step will make your barometer more sensitive to changes in atmospheric pressure.

4. Cut off the open end of the balloon. Next, stretch the balloon over the open end of the coffee can. Then, attach the balloon to the can with masking tape or a rubber band.

| Under Pressure! *continued*

5. Cut one end of the straw at an angle to make a pointer.

6. Place the straw on the stretched balloon so that the pointer is directed away from the center of the balloon. Five centimeters of the end of the straw should extend past the edge of the can. Tape the straw to the balloon as shown in the illustration below.

7. Tape the index card to the side of the can as shown in the illustration at right. Congratulations! You have just made a barometer!

8. Now, use your barometer to collect and record information about air pressure. Place the barometer outside for 3 or 4 days. On each day, mark on the index card where the tip of the straw points.

ANALYZE THE RESULTS

9. **Explaining Events** What atmospheric factors affect how your barometer works? Explain your answer.

____

____

____

____

____

10. **Recognizing Patterns** What does it mean when the straw moves up?

____

____

____

11. **Recognizing Patterns** What does it mean when the straw moves down?

____

____

____

____

❙ Under Pressure! *continued*

DRAW CONCLUSIONS

12. Applying Conclusions Compare your results with the barometric pressures listed in your local newspaper. What kind of weather is associated with high pressure? What kind of weather is associated with low pressure?

____

____

____

____

13. Evaluating Results Does the barometer you built support your hypothesis? Explain your answer.

____

____

____

____

BIG IDEA QUESTION

14. Drawing Conclusions Why is it important to be aware of barometric pressure? How can pressure help scientists to predict weather?

____

____

____

____

Skills Practice Lab

Under Pressure!

Imagine that you are planning a picnic with your friends. The temperature this afternoon should be in the low 80s, and the skies should be sunny. These weather conditions sound quite comfortable. But you know that weather can change when air pressure changes. You can use a barometer to keep track of air-pressure changes. In this activity, you will build your own barometer and will discover what this tool can tell you.

OBJECTIVES

Predict how changes in air pressure affect weather.

Build a barometer to test your hypothesis.

MATERIALS

- balloon
- can, coffee, large, empty, 10 cm in diameter
- card, index

- scissors
- straw, drinking
- tape, masking, or rubber band

SAFETY INFORMATION 

Using Scientific Methods

ASK A QUESTION

1. How can I use a barometer to detect changes in the weather?

FORM A HYPOTHESIS

2. Write a hypothesis to answer the question above.

____

____

____

▌Under Pressure! *continued*

TEST THE HYPOTHESIS

3. Stretch the balloon a few times. Then, blow up the balloon, and let the air out. This step will make your barometer more sensitive to changes in atmospheric pressure.

4. Cut off the open end of the balloon. Next, stretch the balloon over the open end of the coffee can. Then, attach the balloon to the can with masking tape or a rubber band.

5. Cut one end of the straw at an angle to make a pointer.

6. Place the straw on the stretched balloon so that the pointer is directed away from the center of the balloon. Five centimeters of the end of the straw should extend past the edge of the can. Tape the straw to the balloon as shown in the illustration at right.

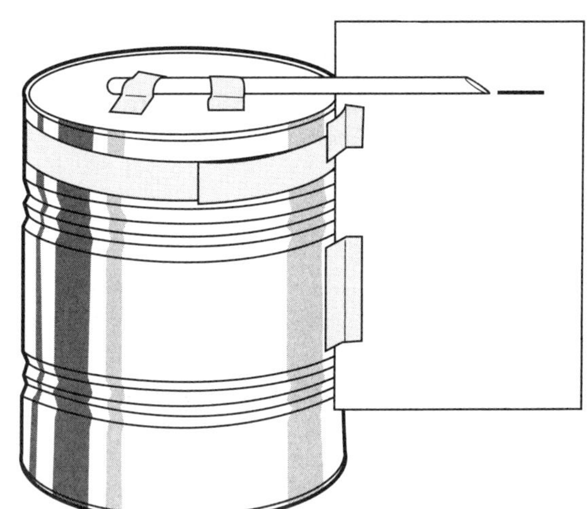

7. Tape the index card to the side of the can as shown in the illustration at right. Congratulations! You have just made a barometer!

8. Now, use your barometer to collect and record information about air pressure. Place the barometer outside for 3 or 4 days. On each day, mark on the index card where the tip of the straw points.

ANALYZE THE RESULTS

9. Explaining Events What factors affect how your barometer works? Explain your answer.

____

____

____

10. Recognizing Patterns How does your barometer behave when the air pressure is high?

____

____

____

____

11. Recognizing Patterns How does your barometer behave when the air pressure is low?

____

____

____

DRAW CONCLUSIONS

12. Applying Conclusions Compare your results with the barometric pressures listed in your local newspaper. What kinds of weather are associated with high and low pressures?

____

____

____

13. Evaluating Results Does the barometer you built support your hypothesis? Why or why not?

____

____

____

BIG IDEA QUESTION

14. Making Inferences Why is it important to be aware of barometric pressure? How can pressure help scientists to predict weather?

____

____

____

____

____

DATASHEET

Preparing for an Oral Report

Teacher Notes

This activity has students use data collected during a Quick Lab to prepare an oral report (covers standard 6.7.d). Encourage students to place their sample collection cards in a variety of locations, such as near a heavily traveled street, under a large tree, or in an out-of-the-way corner. Students who research on the Internet will find *particulate air pollution* to be useful keywords.

Preparing for an Oral Report

INVESTIGATION AND EXPERIMENTATION

6.7.d Communicate the steps and results from an investigation in written reports and oral presentations.

TUTORIAL

Communicating your results is an important part of any scientific investigation. Sometimes, scientists give oral reports. Preparation is the key to giving a good oral report. The steps outlined below will help you prepare for an oral report.

1. First, research the topic you will be discussing. Ask your teacher what sources you should use for the assignment.

2. To guide your research, make a list of topics you want to discuss in your presentation. You can update this list as you continue to gather information. Remember to take good notes!

3. Think about the order in which you will discuss the information. Create an outline that lists the order of topics.

4. Think about your introduction and conclusion. Your introduction should get your audience's attention. Your conclusion should review the information you have discussed.

5. Practice giving your presentation in front of family members or a mirror.

6. Use your outline as a guide when you give your presentation. Look at your audience, stand straight, stay still, and speak clearly.

Preparing for an Oral Report *continued*

YOU TRY IT!
Procedure

Some substances that can be air pollutants include dust and smoke particles, pollen, mold spores, and waste gases. If these tiny particles remain suspended in the air for long periods of time, they are called *particulates*.

You will prepare and present an oral report about the particulates you collected when you performed the Quick Lab entitled "Collecting Air-Pollution Particles." Use the library or the Internet to research common particulates, their sources, and factors that affect the amount of particulates in the air. Your presentation should answer the following questions:

1. What are some common particulates? Are these particulates common where you live? Did you collect any of these particulates during your investigation?

____

____

____

____

____

2. What are possible sources of common particulates where you live?

____

____

____

____

____

3. How does wind affect the amount of particulates in the air?

____

____

____

____

Answer Key

Directed Reading A

SECTION: CHARACTERISTICS OF THE ATMOSPHERE

1. C
2. B
3. C
4. C
5. B
6. B
7. D
8. B
9. D
10. C
11. C
12. B
13. B
14. A

SECTION: ATMOSPHERIC HEATING

1. C
2. B
3. C
4. A
5. A
6. B
7. D
8. B
9. D
10. B
11. D
12. C
13. C

SECTION: AIR MOVEMENT AND WIND

1. A
2. C
3. D
4. A
5. B
6. B
7. local wind
8. sea breeze
9. land breeze
10. valley breeze
11. mountain breeze

SECTION: THE AIR WE BREATHE

1. B
2. B
3. C
4. D
5. A
6. A
7. D
8. D
9. B
10. D
11. B
12. C
13. D
14. C
15. A
16. B
17. D
18. A
19. B

Directed Reading B

SECTION: CHARACTERISTICS OF THE ATMOSPHERE

1. B
2. C
3. A
4. C
5. B
6. C
7. air pressure
8. Answers may vary. Sample answer: As you move farther from Earth's surface, there are fewer gas molecules above. As altitude increases, air pressure decreases.
9. Answers may vary. Sample answer: Some parts of the atmosphere are warmer because they contain a high percentage of gases that absorb solar energy. Parts of the atmosphere that are cooler contain less of these gases.
10. B
11. C
12. A
13. D
14. The atmosphere is divided into four layers based on temperature changes.

15. Answers may vary. Sample answer: Temperature falls as altitude increases because air gets thinner and does not absorb solar radiation that can warm it.

16. Answers may vary. Sample answer: The ozone in the stratosphere and the nitrogen and oxygen in the thermosphere absorb ultraviolet radiation from the sun, which warms the air.

SECTION: ATMOSPHERIC HEATING

1. C
2. D
3. B
4. C
5. the electromagnetic spectrum
6. absorbed
7. visible light
8. Heat is transferred through a material by direct physical contact between particles.
9. Answers may vary. Sample answer: Atoms or molecules with more kinetic energy transfer energy to atoms or molecules with less kinetic energy. Hot objects have atoms with greater average kinetic energy than do the atoms in cold objects; therefore, heat is transferred from a hot object to a cold object.
10. A
11. B
12. Answers may vary. Sample answer: Atmospheric gases, such as water vapor and carbon dioxide, absorb thermal energy and radiate it back to earth. The gases function like the glass walls and roof of a greenhouse, allowing solar energy to enter and preventing thermal energy from escaping.
13. radiation balance
14. global warming
15. Greenhouse gases are gases that absorb thermal energy in the atmosphere.
16. Answers may vary. Sample answer: the burning of fossil fuels and deforestation

SECTION: AIR MOVEMENT AND WIND

1. C
2. B

3. C
4. B
5. D
6. Coriolis effect
7. C
8. D
9. A
10. B
11. Answers may vary. Sample answer: Local geographic features, such as shorelines or mountains, can produce temperature differences that cause local winds.
12. C
13. B
14. A
15. D

SECTION: THE AIR WE BREATHE

1. air pollution
2. Answers may vary. Sample answer: volcanic ash, pollen, forest fire smoke
3. Answers may vary. Sample answer: When primary pollutants react with other primary pollutants or with naturally occurring substances, secondary pollutants form.
4. Answers may vary. Sample answer: ozone, smog
5. Answers may vary. Sample answer: Ozone can damage the lungs.
6. Smog forms when sunlight reacts with ozone and vehicle exhaust.
7. Los Angeles is almost completely surrounded by mountains that trap pollutants and contribute to smog formation.
8. A
9. Answers may vary. Sample answer: burning fossil fuel, production of chemicals
10. Answers may vary. Sample answer: ventilation, limiting the use of chemical solvents and cleaners
11. C
12. B
13. B
14. Answers may vary. Sample answer: Eastern Europe, the northeastern United States, eastern Canada
15. acid shock
16. The ozone hole allows more ultraviolet (UV) radiation to reach Earth's surface.

17. 60 to 120 years

18. UV radiation damages genes and causes skin cancer.

19. Answers may vary. Sample answer: Air pollution can cause coughing, headaches, an increase in asthma-related problems, and lung cancer.

20. C

21. Answers may vary. Sample answer: Industries may use pollution control devices, such as scrubbers; industrial plants may burn fuel more efficiently so fewer pollutants are released.

22. Answers may vary. Sample answer: The EPA sets allowances for the amount of a pollutant that companies can release, and it fines companies that exceed their allowances.

23. Answers may vary. Sample answer: Car manufacturers place catalytic converters in cars to reduce pollution from exhaust; they make cars that run on hydrogen or natural gas; they make hybrid cars that use both gasoline and electric power.

24. Answers may vary. Sample answer: People can carpool, use public transportation, bike, or walk.

Vocabulary and Section Summary A

SECTION: CHARACTERISTICS OF THE ATMOSPHERE

1. atmosphere: a mixture of gases that surrounds a planet or moon

2. air pressure: the measure of the force with which air molecules push on a surface

3. troposphere: the lowest layer of the atmosphere, in which temperature decreases at a constant rate as altitude increases

4. stratosphere: the layer of the atmosphere that is above the troposphere and in which temperature increases as altitude increases

5. mesosphere: the layer of the atmosphere between the stratosphere and the thermosphere and in which temperature decreases as altitude increases

6. thermosphere: the uppermost layer of the atmosphere, in which temperature increases as altitude increases

SECTION: ATMOSPHERIC HEATING

1. radiation: the transfer of energy as electromagnetic waves

2. electromagnetic spectrum: all of the frequencies or wavelengths of electromagnetic radiation

3. conduction: the transfer of energy as heat through a material

4. convection: the movement of matter due to differences in density; the transfer of energy due to the movement of matter

5. convection current: any movement of matter that results from differences in density; may be vertical, circular, or cyclical

6. greenhouse effect: the warming of the surface and lower atmosphere of Earth that occurs when water vapor, carbon dioxide, and other gases absorb and reradiate thermal energy

SECTION: AIR MOVEMENT AND WIND

1. wind: the movement of air caused by differences in air pressure

2. Coriolis effect: the curving of the path of a moving object from an otherwise straight path due to Earth's rotation

SECTION: THE AIR WE BREATHE

1. air pollution: the contamination of the atmosphere by the introduction of pollutants from human and natural sources

2. acid precipitation: rain, sleet, or snow that contains a high concentration of acids

Vocabulary and Section Summary B

SECTION: CHARACTERISTICS OF THE ATMOSPHERE

Across

2. mesosphere

5. troposphere

6. thermosphere

Down
1. atmosphere
3. stratosphere
4. air pressure

SECTION: ATMOSPHERIC HEATING
1. convection
2. electromagnetic spectrum
3. conduction
4. radiation
5. convection current
6. greenhouse effect

SECTION: AIR MOVEMENT AND WIND
1. wind
2. Coriolis effect
3. atmosphere
4. global winds
5. local winds
6. conduction
7. convection
8. radiation

Q	T	F	H	Y	Q	K	M	I	D	J	L	B	B	H	Y	D	L	C	L
O	A	J	Q	U	H	A	M	X	F	U	Y	Q	V	Q	T	J	O	N	O
I	N	U	I	W	Y	I	G	M	J	R	V	M	N	D	C	R	U	N	C
L	Z	T	K	Z	K	N	H	D	B	T	J	T	S	J	I	Z	O	U	A
H	Z	J	H	Z	H	M	E	U	R	D	V	D	F	O	J	I	D	U	L
R	T	U	H	Y	R	V	D	V	Z	E	I	R	L	R	T	P	K	W	W
R	X	S	T	H	S	N	P	U	H	Z	R	I	X	C	W	D	R	T	I
G	L	O	B	A	L	W	I	N	D	S	S	E	U	D	G	I	H	K	N
Z	H	R	B	W	H	S	Y	Q	E	E	V	D	H	M	K	T	N	R	D
K	V	H	W	X	D	L	P	E	F	A	N	R	I	P	R	V	K	D	S
B	M	N	Y	V	X	Q	W	E	C	O	P	I	V	Y	S	O	J	G	I
S	Y	I	U	O	W	O	C	G	C	S	B	B	Z	M	N	O	J	F	L
B	B	M	E	Q	E	T	R	Y	I	G	B	F	N	U	K	O	M	J	V
X	R	Q	R	V	S	P	Y	T	V	M	Q	B	F	V	O	K	U	T	K
R	A	D	I	A	T	I	O	N	G	S	Y	X	S	G	W	T	R	A	A
C	B	U	E	W	T	Z	X	R	L	Z	B	C	W	V	M	H	H	A	R
N	E	M	A	N	F	P	L	L	J	Q	B	O	M	L	M	W	W	H	A
J	K	Z	G	Z	W	H	Y	L	C	P	Z	S	J	T	V	G	R	H	C
E	H	S	F	S	Y	F	U	N	O	I	T	C	E	V	N	O	C	D	L

SECTION: THE AIR WE BREATHE
1. CFC
2. primary pollutants
3. scrubber
4. air pollution
5. acid precipitation
6. ventilation
7. acidification
8. secondary pollutants
9. acid shock
10. Clean Earth

Reinforcement

EARTH'S AMAZING ATMOSPHERE
1. A
2. C
3. D
4. B

Critical Thinking

1. No, it seems like an unreliable source. Many supermarket tabloids like this one are unreliable sources of scientific facts. Also, the scientist whose work is presented has no formal qualifications, and the fact that his last article is about a time machine makes him a doubtful source.
2. It will catch solid particles in the air. These include dust, ash, and smoke.
3. The particles cannot build up forever inside the machine. Eventually, the machine will fill up.
4. Answers may vary. Sample answer: He incorrectly assumes that all air pollution is composed of solid particles. He also assumes that all the air pollution in the world can be sucked up by a single machine. He assumes that the GBG5K will be able to distinguish between things we want in the air—such as insects, birds, and pollen—and things we do not want in the air, such as pollution.
5. Answers may vary. Sample answer: I would suggest that he redesign his machine to remove gaseous pollutants as well as unwanted solid particles.

SciLinks Activity

1. Answers may vary. Sample answer: Ozone can be good or bad depending on where it is found. Good ozone occurs naturally 10 to 30 miles above Earth's surface, where it forms a protective layer that shields us from the sun's harmful UV radiation. Bad ozone occurs near Earth's surface. It is formed when pollutants emitted by cars, power plants, industrial boilers, refineries, chemical plants, and other sources react chemically in the presence of sunlight.

2. Answers may vary. Sample answer: Ozone can irritate the respiratory system. It might cause a person to cough, feel an irritation in the throat, and feel an uncomfortable sensation in the chest. Ozone can reduce lung function (the volume of air that can be drawn in with a full breath and the speed at which air is blown out). Ozone can aggravate asthma. When ozone levels are high, asthmatics might have serious attacks. Some studies suggest that ozone may reduce the immune system's ability to fight disease.

3. Answers may vary. Sample answer: Ozone damage can occur without any noticeable signs. Often, people exposed to ozone experience recognizable symptoms, including coughing, irritation in the airways, rapid or shallow breathing, and discomfort when breathing or general discomfort in the chest. People with asthma may experience asthma attacks.

4. Answers may vary. Sample answer: You can reduce ozone levels by walking, biking, carpooling, or using public transportation as an alternative to traveling by car. Keeping a car well tuned also reduces emissions.

Section Review

SECTION: CHARACTERISTICS OF THE ATMOSPHERE

1. The temperature of the different layers of the atmosphere varies because the gases at different altitudes absorb solar radiation differently.

2. Air pressure is the measure of the force with which air molecules push on a surface. It is equal to the weight of the air directly above that point.

3. Air pressure decreases as altitude increases because the weight of the overlying atmosphere decreases.

4. nitrogen and oxygen; 99%

5. 0.9%

6. Air density is lower at higher altitudes and helicopters need air to provide lift. At altitudes higher than 6,000 m, air density is too low for the helicopter to achieve lift.

7. Air pressure is highest at Earth's surface and decreases as you move away from Earth's surface. The layers of the atmosphere in order of decreasing air pressure are troposphere, stratosphere, mesosphere, and thermosphere. Temperature varies in the atmosphere depending on how gases absorb solar radiation. In the troposphere, temperature decreases as altitude increases. In the stratosphere, temperature increases as altitude increases. In the mesosphere, temperature decreases as altitude increases. In the thermosphere, temperature increases as altitude increases.

8. $50 \text{ km} - 12 \text{ km} = 38 \text{ km}$

SECTION: ATMOSPHERIC HEATING

1. Sample answer: Conduction is the transfer of thermal energy as heat through a material. Radiation is the transfer of energy by electromagnetic waves. Convection is the transfer of energy by circulation in a fluid. Global warming is a gradual increase in average global temperature.

2. The sun is the major source of energy for many processes at Earth's surface.

3. Radiation is the transfer of energy as electromagnetic waves through matter or through space.

4. Energy is transferred in the atmosphere by radiation, by conduction, and by convection.

5. Radiation is different from conduction and convection in that radiation can transfer energy through empty space.

6. The electromagnetic spectrum contains all of the frequencies or wavelengths of electromagnetic radiation, or all of the kinds of electromagnetic waves. Electromagnetic radiation reaches Earth from the sun by radiation, or by traveling as waves through space.

7. In conduction, energy is transferred by the direct contact between particles. Conduction happens in solids and in fluids and involves no flow of matter. In convection, energy is transferred by the movement of matter. Convection happens only in fluids.

8. Convection begins when energy is transferred from one substance to another by conduction.

9. The greenhouse effect is Earth's natural heating process by which gases in the atmosphere absorb reradiated energy, which heats the atmosphere. Global warming is a rise in average global temperature and may be caused by an increase in the greenhouse effect.

10. $(22.8°C + 21.7°C + 12.5°C + 18.6°C + 25.7°C + 27.7°C + 27.8°C) \div 7 = 22.4°C$

11. Sample answer: If the amount of incoming radiation from the sun were not uniform, the processes of conduction and convection would not be as predictable on Earth's surface.

SECTION: AIR MOVEMENT AND WIND

1. Sample answer: Wind is the movement of air. The Coriolis effect is the deflection of a moving object due to Earth's rotation.

2. The uneven heating of Earth's surface produces pressure belts. These belts form where warm air is pushed upward near the equator and at 60°N and 60°S latitude or where cold air sinks near the poles and at 30°N and 30°S latitude.

3. Winds are caused by the unequal heating of Earth's surface, which causes pressure differences.

4. The rotation of Earth on its axis causes the Coriolis effect.

5. The Coriolis effect causes winds to be deflected to the east or west depending on the direction that the winds are traveling in each hemisphere. Because of the Coriolis effect, winds in the Northern Hemisphere curve to the right, and winds in the Southern Hemisphere curve to the left.

6. Global winds are caused by the uneven heating of Earth's surface due to the amount of direct sunlight that reaches the ground at different latitudes. Local winds are caused by the uneven heating of Earth's surface due to the way that different materials absorb and release energy.

7. Sea and land breezes form as a result of the different rates at which land and water heat up and cool down. Mountain and valley breezes form as a result of the topography and the direction in which the air flows when it heats up and cools down.

8. Because unequal heating of Earth's surface causes winds, there would probably not be winds near Earth's surface if Earth's surface were the same temperature everywhere.

9. During the day, a sea breeze is caused by cooler air over the water moving toward the land. Walking into the sea breeze would lead you to the ocean.

10. In the Northern Hemisphere, the westerlies appear to be deflected to the northeast by the Coriolis effect. In contrast, the trade winds blow in a southerly direction, so they appear to be deflected to the southwest.

SECTION: THE AIR WE BREATHE

1. Acid shock

2. Acid precipitation

3. Vehicle exhaust reacts with sunlight and ozone to create smog.

4. Primary pollutants are pollutants that are put into the air directly by human or natural activity. Secondary pollutants form when primary pollutants react with other substances.

5. Sample answer: household cleaners, paint products, dirty air-conditioning vents, furniture, and stoves

6. When fossil fuels are burned, they release pollutants, such as sulfur dioxide and nitrogen oxide. These chemicals react with water in the atmosphere to cause acid precipitation. They can also react with sunlight and ozone to form smog.

7. Answers may vary but should include five of the following: headache; nausea; irritation of the eyes, nose, and throat; coughing; upper respiratory infections; and worsening of asthma and emphysema.

8. No, some air pollutants, such as volcanic ash and gases, pet dander, and plant pollen, come from natural sources.

9. Answers may vary. Accept all reasonable answers.
10. Sample answer: Health problems can be affected by many factors, including age, nutrition, background, and genetics.
11. Sample answer: Western New York and Pennsylvania have the most acidic precipitation. These areas have a lot of industry that produces a lot of pollution. That pollution leads to acid precipitation.
12. Sample answer: Buffalo is closer to upwind pollution sources than Boston is.
13. Sample answer: Air pollution can move all over the globe as a result of global air circulation by convection currents.

Chapter Review

1. transfer
2. Conduction is the transfer of heat between objects that are touching; it does not involve the movement of matter. Convection is the transfer of heat by the movement of matter.
3. Radiation is the transfer of energy through matter or space as waves. The electromagnetic spectrum is all of the wavelengths of electromagnetic radiation.
4. Wind is the movement of air due to differences in pressure that are caused by the uneven heating of Earth's surface by the sun. Air pressure is the force with which air molecules push on a surface.
5. C
6. A
7. B
8. primary: car exhaust, smoke from a factory, fumes from burning plastic; secondary: acid precipitation
9. Air pressure decreases as altitude increases because less air pushes down from above.
10. When air is heated, it expands and becomes less dense. When colder, denser air moves into an area, it pushes the warmer, less dense air upward and out of the way.
11. Temperature changes in the atmosphere result from how different gases absorb solar radiation.

12. Secondary pollutants are pollutants that form when primary pollutants react with other primary pollutants or with naturally occurring substances or energy. Smog and acid precipitation are secondary pollutants.
13. The electromagnetic spectrum is the range of all wavelengths of electromagnetic radiation. Visible light represents only a very small amount of electromagnetic radiation.
14. Incoming solar radiation is more direct at the equator than at the poles. As a result, solar energy strikes Earth's surface more directly at the equator, and temperatures there are high. An area of low pressure forms at the equator. At the poles, the air is cold and dense because it absorbs little heat. Therefore, the air sinks and forms an area of high pressure. These pressure differences combine with the Coriolis effect and convection cells to form the global wind systems.
15. developed countries
16. Eastern Europe
17. Answers may vary but should cover all requirements stated in the question.
18. Answers may vary but should cover all requirements stated in the question.
19. An answer to this exercise can be found at the end of the Teacher Edition.
20. The greenhouse effect increases the planet's temperature.
21. Sample answer: I think the Coriolis effect would be twice as noticeable if the planet spun twice as fast.
22. Global winds are the result of the combination of the Coriolis effect and convection cells which occur at every 30° of latitude. The convection cells result from the uneven heating of Earth's surface.
23. Local winds form because of the unequal heating of Earth's surface that results from differences in how land and water absorb and release heat. Global winds result from differences in the amount of solar radiation received by different latitudes.
24. By radiation, energy can travel through space, whereas in conduction and convection, energy travels through matter.

25. Air and stratospheric ozone are both renewable resources.

26. Energy from the sun travels to Earth through space as electromagnetic waves.

27. Visible and nonvisible light differ in their wavelengths and in the ways in which they are absorbed by different gases in the atmosphere.

28. increasing

29. Sample answer: Humans could be adding carbon dioxide to the air by burning fossil fuels.

30. Sample answer: If the trend continues, the amount of carbon dioxide in the atmosphere will increase. Humans can affect this result by burning fewer fossil fuels.

31. (12 km/h + 20 km/h + 11 km/h + 6 km/h + 8 km/h + 19 km/h + 17 km/h) ÷ 7 = 13 km/h

32. Sample answer: The two suns could increase the amount of solar radiation received by the planet's surface near the equator. This increase would cause a more drastically uneven heating of the planet's surface. The larger difference in air pressure would cause the winds to be stronger.

Section Quizzes

SECTION: CHARACTERISTICS OF THE ATMOSPHERE

1. B
2. C
3. B
4. D
5. C
6. D
7. B
8. A

SECTION: ATMOSPHERIC HEATING

1. C
2. B
3. A
4. C
5. B
6. B
7. C

SECTION: AIR MOVEMENT AND WIND

1. A
2. A
3. C
4. C
5. B
6. B
7. A
8. D
9. A

SECTION: THE AIR WE BREATHE

1. C
2. B
3. C
4. C
5. B
6. C
7. C
8. B
9. D
10. A

Chapter Test A

1. C
2. D
3. B
4. B
5. C
6. B
7. B
8. A
9. D
10. C
11. B
12. A
13. B
14. C
15. D
16. A
17. radiation
18. conduction
19. global warming
20. convection current

Chapter Test B

1. C
2. D
3. A
4. B
5. A
6. D
7. A
8. A
9. C
10. A
11. B
12. B
13. C
14. A
15. D
16. F
17. E
18. H
19. I
20. G
21. D
22. B
23. E
24. A
25. C

Chapter Test C

1. conduction
2. secondary
3. westerlies
4. greenhouse effect
5. Coriolis effect
6. B
7. D
8. A
9. Answers may vary. Sample answer: At night, the air over land cools more quickly than the air over water, creating an area of high pressure. Warmer air over water creates an area of low pressure. Because air always moves from areas of high pressure to areas of low pressure, air moves from the land to the ocean and causes a land breeze.
10. Answers may vary. Sample answer: At nightfall, air along mountain slopes cools. This cool air moves down the slopes into the valley, producing a mountain breeze.
11. Answers may vary. Sample answer: Both gases in Earth's atmosphere and the glass covering of a greenhouse allow solar energy to pass through them but stop energy from escaping to the outside.
12. Answers may vary. Sample answer: The temperature in the troposphere would drop. Normally, the sun's radiation reaches the troposphere, and the energy from radiation is transferred into heat. If the stratosphere were covered in a layer of volcanic dust, less of the sun's radiation would reach Earth's surface, and less heat would be produced.
13. Answers may vary. Sample answer: If more fossil fuels were burned, more sulfuric acid and nitric acid would form, increasing the acidity of precipitation. In aquatic ecosystems, high acidity would threaten plant and animal life. On land, forests and other plant life would be threatened as soil acidity increased.
14. Answers may vary. Sample answer: The ozone layer protects Earth from harmful ultraviolet radiation. UV radiation damages genes and can cause skin cancer. Without the ozone layer, most life on Earth would be threatened with extinction.
15. Answers may vary. Sample answer: The graph shows a steady increase in CO_2 concentration. Some scientists believe that an increase in carbon dioxide levels has led to an increase in global temperatures. Higher global temperatures could cause serious climatic and weather changes.
16. **a.** thermosphere, **b.** density, **c.** heat, **d.** temperature, **e.** particles in motion, **f.** nitrogen and oxygen, **g.** solar energy

Performance-Based Assessment

Data Table		
Time (min)	Covered (°C)	Uncovered (°C)
0	21	21
2	33	31
4	37	33
6	39.5	34.5
8	42	36
10	43.5	37
12	41	35
14	39	34
16	37	33
18	36	32
20	34.3	31

5. The covered terrarium heated up at a much faster rate than the uncovered terrarium.

6. With the cover, the terrarium loses heat much more slowly.

7. Answers may vary. Sample answer: Yes, wet sand holds heat much longer than dry sand does. Water retains heat.

8. Answers may vary. Sample answer: The covered terrarium is a better model. Earth's atmosphere acts like a "cover" over Earth. Like the cover on the terrarium, Earth's atmosphere helps trap heat.

9. Answers may vary. Sample answer: Living things need heat to survive. That heat becomes available from the solar energy that arrives on Earth and must somehow be changed to heat and retained to make Earth livable. This is what Earth's atmosphere does. Solar energy (light in the experiment) is changed to heat by atmospheric gases, such as water vapor (water in the experiment). Much of this energy is reradiated into space as heat, but enough remains to sustain life.

Explore Activity

DATASHEET A

3. Answers may vary. Accept all reasonable answers.

5. Answers may vary. Accept all reasonable answers.

6. Data should show that one thermometer—the one onto which light reflected—indicated a temperature increase.

7. The energy that caused the temperature change came from the sun. The energy was reflected to one thermometer by the mirror. The energy reached the other thermometer by bouncing off the ground or air.

8. Sample answer: If additional mirrors were used to focus light on the thermometer, sunlight from a larger area would be focused on the thermometer, so the temperature increase would be greater.

DATASHEET B

6. Data should show that one thermometer—the one onto which light reflected—indicated a temperature increase.

7. Energy that caused the temperature change came from the sun. It was reflected to one thermometer by the mirror. It reached the other thermometer by bouncing off the ground or air.

8. Sample answer: If additional mirrors were used to focus light on the thermometer, sunlight from a larger area would be focused on the thermometer, so the temperature increase would be greater.

DATASHEET C

6. Data should show that one thermometer—the one onto which light reflected—indicated a temperature increase.

7. Energy that caused the temperature change came from the sun. It was reflected to one thermometer by the mirror. It reached the other thermometer by bouncing off the ground or air.

8. Sample answer: If additional mirrors were used to focus light on the thermometer, sunlight from a larger area would be focused on the thermometer, so the temperature increase would be greater. If additional mirrors were positioned to reflect light away from the thermometer, less energy would reach the thermometer and the temperature would not increase.

Quick Lab: Modeling Air Pressure

DATASHEET A

4. Sample answer: No, the water is held in the cup by the index card.

5. air pressure; greater

DATASHEET B

5. Sample answer: Something has to be holding the paper up; it must be air pressure.

DATASHEET C

4. Sample answer: The water is held in the cup by the index card.

5. Sample answer: Air pressure; The force of air pressure holding the index card up must be greater than the force of gravity pulling down on the water.

Quick Lab: Modeling Air Movement by Convection

DATASHEET A

5. Answers may vary due to slight differences in temperature. Accept all reasonable answers.

8. The temperature of the air near the ice cubes is cool, and the incense smoke moves downward because cooler air is denser. The temperature of the air near the lamp is warmer, and the incense smoke swirls near the top of the aquarium because warmer air is less dense.

9. This activity models convection in the atmosphere because air sinks in the cold areas and rises in the warm areas. This movement produces a circular pattern called a *convection current*.

DATASHEET B

8. The temperature of the air near the ice cubes is cool, and the incense smoke moves downward. The temperature of the air near the lamp is warmer, and the incense smoke swirls near the top of the aquarium.

9. This activity models convection in the atmosphere because air sinks in the cold areas and rises in the warm areas. This movement produces a circular pattern called a *convection current*.

DATASHEET C

5. Answers may vary due to slight differences in temperature. Accept all reasonable answers.

8. The temperature of the air near the ice cubes is cool, and the incense smoke moves downward. The temperature of the air near the lamp is warmer, and the incense smoke swirls near the top of the aquarium.

9. This activity models convection in the atmosphere because air sinks in the cold areas and rises in the warm areas. This movement produces a circular pattern called a *convection current*.

Quick Lab: Investigating the Coriolis Effect

DATASHEET A

3. Students should observe that the water drops trace a path that spirals off the disk.

4. Sample answer: Near the pencil at the center of the disk, the cardboard disk spins more slowly than the edge of the cardboard disk does. As the water moves from the center of the disk toward the edge, the water moves to the side more slowly than the spinning cardboard does. As a result, the water is deflected to the side in a curved path. This is similar to the Coriolis effect of Earth's rotation on ocean currents and global winds.

DATASHEET B

3. Students should observe that the water drops trace a path that spirals off the disk.

4. Sample answer: The spinning disk represents Earth. The pencil represents Earth's rotational axis. Near the pencil at the center of the disk, the cardboard disk spins more slowly than the edge of the cardboard disk does. As the water moves from the center of the disk toward the edge, the water moves to the side more slowly than the spinning cardboard does. As a result, the water is deflected to the side in a curved path.

DATASHEET C

3. Students should observe that the water drops trace a path that spirals off the disk.

4. Sample answers: The spinning disk represents Earth. The pencil represents Earth's rotational axis. Near the pencil at the center of the disk, the cardboard disk spins more slowly than the edge of the cardboard disk does. As the water moves from the center of the disk toward the edge, the water moves to the side more slowly than the spinning cardboard does. As a result, the water is deflected to the side in a curved path. The curve of the path from the water should go in the opposite direction. This is what occurs with the Coriolis effect between the Northern and Southern Hemispheres of Earth.

Quick Lab: Collecting Air-Pollution Particles

DATASHEET A

4. Students' observations may vary. However, cards placed in or near high-traffic areas and directly under trees are likely to have the highest number of particulates.

5. Sample answer: Automobile traffic might contribute a large amount of particulate matter to the air. During certain times of the year, trees might also contribute significant particulates.

DATASHEET B

4. Students' observations may vary. However, cards placed in or near high-traffic areas and directly under trees are likely to have the highest number of particulates.

5. Students might hypothesize that automobile traffic contributes a large amount of particulate matter to the air. During certain times of the year, trees will also contribute significant particulates.

DATASHEET C

4. Students' observations may vary. However, cards placed in or near high-traffic areas and directly under trees are likely to have the highest number of particulates.

5. Students might hypothesize that automobile traffic contributes a large amount of particulate matter to the air. During certain times of the year, trees will also contribute significant particulates. Answers may vary. Accept any reasonable answers. Sample answer: Primary pollutants included dust, pollen, smoke, or vehicle exhaust. With the method being used, it may not be possible to collect secondary pollutants, such as ozone.

Chapter Lab

DATASHEET A

2. atmospheric pressure; up; down

9. A change in air pressure will affect the barometer. Temperature changes may also affect the barometer.

10. Upward movement of the straw indicates that the atmospheric pressure is increasing.

11. Downward movement of the straw indicates that the atmospheric pressure is decreasing.

12. Clear, dry days are associated with high pressure.
Cloudy, rainy, or humid days are associated with low pressure.

13. Answers may vary depending on the hypothesis. Students should explain how their barometer was affected by atmospheric pressure and why the experiment supported or disproved their hypothesis.

14. Sample answer: Scientists can use atmospheric pressure to predict the weather by measuring it with a barometer to find out if the pressure is high or low. Sample answer: Changes in air pressure can help predict when weather will be clear and sunny or cloudy and humid. This information helps people know what kind of clothing they need to wear when they go outside. Also, the changes in air pressure can warn of storms, such as hurricanes. In such cases, knowing the air pressure would help people plan for, or avoid, dangerous weather conditions.

DATASHEET B

9. A change in air pressure will affect the barometer. Temperature changes may also affect the barometer.
10. Upward movement of the straw indicates that the atmospheric pressure is increasing. Air pressure is pushing on the balloon, which causes the pointer to rise.
11. Downward movement of the straw indicates that the atmospheric pressure is decreasing. Less air pressure causes the pointer to dip downward.
12. Clear, dry days are associated with high pressure. Cloudy, rainy, or humid days are associated with low pressure. A sudden drop in air pressure usually indicates that a storm is on the way.
13. Answers may vary depending on the hypothesis. Students should explain how their barometer was affected by atmospheric pressure and why the experiment supported or disproved their hypothesis.
14. Changes in air pressure can help predict when weather will be clear and sunny or cloudy and humid. This information helps people know what kind of clothing they need to wear when they go outside. Also, changes in air pressure can warn of storms, such as hurricanes. In such cases, knowing the air pressure would help people plan for, or avoid, dangerous weather conditions.

DATASHEET C

2. Answers may vary. Accept any reasonable answer.

9. A change in air pressure will affect the barometer. Temperature changes may also affect the barometer.
10. Upward movement of the straw indicates that the atmospheric pressure is increasing. Air pressure is pushing on the balloon, which causes the pointer to rise.
11. Downward movement of the straw indicates that the atmospheric pressure is decreasing. Less air pressure causes the pointer to dip downward.
12. Clear, dry days are associated with high pressure. Cloudy, rainy, or humid days are associated with low pressure. A sudden drop in air pressure usually indicates that a storm is on the way.
13. Answers may vary depending on the hypothesis. Students should explain how their barometer was affected by atmospheric pressure and why the experiment supported or disproved their hypothesis.
14. Changes in air pressure can help predict when weather will be clear and sunny or cloudy and humid. This information helps people know what kind of clothing they need to wear when they go outside. Also, changes in air pressure can warn of storms, such as hurricanes. In such cases, knowing the air pressure would help people plan for, or avoid, dangerous weather conditions.

Science Skills Activity

DATASHEET

Oral reports may vary but should include answers to all three of the questions. Common particulates include dust, diesel soot, fly ash, wood smoke, sulfate aerosols, and pollen. Students' answers may vary. Check to ensure that students correctly identify their particulates. Make sure that students assign the appropriate sources of particulates. Wind might bring particulates from far outside the sampling area. Depending on the wind direction, the wind might increase or decrease the quantity of particulates.